Applied Probability and Statistics

continued on back

Markov Chains

A WILEY PUBLICATION IN MATHEMATICAL STATISTICS

Markov Chains Theory and Applications

DEAN L. ISAACSON
Iowa State University

and

RICHARD W. MADSEN
University of Missouri

John Wiley & Sons

New York · London · Sydney · Toronto

Library of Congress Cataloging in Publication Data

Isaacson, Dean L   1941–
  Markov chains, theory and applications.

  (Wiley series in probability and mathematical statistics)
  Bibliography: p.
  Includes index.
  1. Markov processes. I. Madsen, Richard W., 1941– joint author. II. Title.

QA274.7.I78   519.2'33   75-30646
ISBN 0-471-42862-0

Printed in the United States of America

10 9 8 7 6 5 4 3 2 1

To

Kathy, Jill, and Jeff
Carole, Rick, and Pete

Preface

This book deals entirely with the topic of Markov chains, with most of the emphasis being given to discrete time chains. The topic of Markov chains is of theoretical interest to persons in mathematics, probability, statistics, and computer science and of practical interest to persons in areas such as industrial engineering, economics, psychology, and sociology. While the presentation given here is reasonably rigorous mathematically, we have tried to show the practical uses of Markov chains through examples. In fact, Section 6 of Chapter III is devoted entirely to applications, showing how Markov chains have been used in various fields.

Discrete-time stationary Markov chains can be studied and analyzed from various points of view. The classical or traditional approach to Markov chains is presented in Chapter III. The algebraic approach and a computer approach to Markov chains are presented in Chapters IV and VI. In each of these chapters we consider basic questions of reducibility, persistency, periodicity, and existence of a long run distribution. In Chapters III and VI we also consider absorption times, absorption probabilities, and reduction of a stochastic matrix to block form. In addition we discuss some ratio limit theorems in Chapter III.

Chapters V and VII deal with Markov chains that are not of the "discrete-time stationary" variety. In Chapter V the ergodic coefficient is introduced and is used to study questions of ergodicity for nonstationary Markov chains. In Chapter VII some basic properties of continuous-time Markov chains are presented. As an application of these properties, birth and death processes are studied, providing an introduction to some concepts of queuing theory.

This textbook is suitable for a course taught at the senior undergraduate level or first year graduate level. It may be used for a one-semester course or for a one-quarter course if parts of Chapters IV, VI, or VII are deleted.

To get the most out of such a course a student should have some knowledge of elementary probability theory, advanced calculus, and should know enough linear algebra to be able to multiply vectors and matrices. A review of some of the necessary probability theory and certain aspects of analysis that are useful in reading this book are given in Chapter I. Some of the fundamental concepts of linear algebra needed in Chapter IV are presented there. A student with no background in advanced calculus may have difficulty in following some of the proofs given in Chapters II and III, but the main ideas of these chapters can still be grasped by such a student. The material in Chapter V probably will require more mathematical maturity than the material in the other chapters. In this case it is also true that the main ideas can be understood without thoroughly understanding all of the proofs.

Throughout the book important results are illustrated by examples and exercises. We believe that to get the most out of a course of this type a student must work a reasonably large number of exercises. Consequently we have included a variety of exercises at the end of almost every section. Some of the exercises are designed to help the student understand the material presented and others are designed to extend the material presented. Those exercises that give extensions that are used further along in the text are marked by an asterisk (*). Those exercises that are most difficult have been marked (for reasons which the student may find obvious) with a dagger (†).

This book is the product of the work done by many persons other than ourselves. We would like to extend special thanks to Professors H. T. David and B. C. Arnold for their many helpful discussions on the subject matter, to Professors B. Bowerman and V. Sposito for allowing us to include some of their research in Chapters V and VI, and to the many typists who contributed their typing skills and who showed great patience through many revisions of the manuscript.

DEAN L. ISAACSON
RICHARD W. MADSEN

Ames, Iowa
Columbia, Missouri
May 1975

Contents

CHAPTER

Markov Chains

CHAPTER I

Introduction

SECTION 1 RANDOM VARIABLES

A Markov chain is a special type of stochastic process and a stochastic process is a collection of random variables; therefore it is appropriate to begin Chapter I with a discussion of random variables.

In order to use mathematics to describe an experimental phenomenon it is necessary to build a mathematical *model* to describe the phenomenon. The model used need not be exact in all details, but it should adequately describe the most important aspects of the experiment with which the experimenter is concerned. [See Meyer (1970) for a more expanded discussion of models.]

One of the first steps in modeling an experiment is to define the set of all possible outcomes. In defining sets we will use standard notation. In particular, the notation $\{x : x$ satisfies some condition$\}$ is used to denote the set of all x's such that x satisfies the condition stated. Alternatively, a set may be defined by listing all the elements of the set. For example, the set of even integers from 1 to 6 might be given as $S = \{x : x$ is an even integer between 1 and 6 inclusive$\}$ or $S = \{2, 4, 6\}$.

Definition I.1.1. *The set of all possible outcomes of an experiment is called the* sample space *of the experiment. The sample space will be denoted by the symbol Ω, and an arbitrary element of Ω will be denoted by ω.*

Example I.1.1. If a coin is tossed four times, an appropriate sample space is $\Omega = \{\omega : \omega = (x_1, x_2, x_3, x_4)$ where $x_i \in \{H, T\}$ for $i = 1, 2, 3, 4\}$. There are 16 distinct outcomes in this sample space.

The reader will note that the outcomes of this sample space are detailed enough so that the outcome on each individual toss is described. If,

1

however, one is only interested in whether or not the coins either match (i.e., are all heads or all tails) or they do not match, the sample space should contain only these two outcomes, {match, don't match}. The first sample space would generally be considered more desirable since it allows for a more complete description of the experimental outcomes.

Example I.1.2. A coin is tossed repeatedly until a tail appears. The sample space could be $\Omega = \{\omega : \omega = \text{number of tosses required}\}$ or could be $\Omega = \{\omega : \omega = (x_1, x_2, \ldots, x_n)$ where $x_i = H$ if $i < n$ and $x_n = T$, for $n = 1, 2, \ldots\}$. Both of these sample spaces convey the same information.

Example I.1.3. A student in Engineering 101 is given an assignment that is due in two days. If we are interested in the amount of time he spends working on the assignment, then $\Omega = \{\omega : \omega \in [0, 48]$, where ω is given in hours$\}$.

The number of elements in the sample space of Example I.1.1 is finite while the number in Examples I.1.2 and I.1.3 are countably infinite and uncountably infinite, respectively. In this book we restrict our attention to sample spaces that are either finite or countably infinite. Such sample spaces will be called *discrete*.

Each time an experiment is performed, exactly one of the possible outcomes will occur. Usually it is not known *which* of the possible outcomes will occur. The investigator may feel that each outcome has "as good a chance" of occurring as any other, or he may feel that some outcomes have a "better" chance of occurring than others. In order to translate these "feelings" into precise mathematical terms, the investigator can assign probabilities to each outcome (i.e., to each point in the sample space).

Mathematically, one can define a probability measure P on Ω in any way that is consistent with the following conditions: (1) for each $\omega_i \in \Omega$, there is associated a non-negative number $p_i = P(\omega_i)$ and (2) $\sum_{\omega_i \in \Omega} P(\omega_i) = 1$.

[Referring to Example I.1.1, if we felt that each outcome had "as good a chance" of occurring as any other, we could define $P(\omega_i) = 1/16$ for all i. If we felt that the coin was biased in some way, we could assign probabilities appropriately.]

Definition I.1.2. *If Ω is a finite or countably infinite sample space and if P is a probability measure on Ω, then the pair (Ω, P) is called a* probability space.

Definition I.1.3.   *If* (Ω, P) *is a probability space, then an* event *is any subset of* Ω. *If* $E = \{\omega_1, \omega_2, \ldots, \omega_k\}$, *then the probability of* E *is defined as* $P(E) = \sum_{i=1}^{k} P(\omega_i)$. *The empty set is defined to have probability zero. (As a consequence of this definition we see that* $P(\Omega) = 1$.)

The probability space (Ω, P) contains the information necessary to study the probabilistic properties of the experiment. However, the analysis might be complicated by the fact that the description of the sample points might be in a form that is familiar to the experimental scientist but very foreign to the probabilist. (For example, the points in Ω might be described by compounds resulting from a chemical experiment.) In order to alleviate this communications problem, the points in Ω are often mapped into real numbers. In this way a number can be used to represent an experimental outcome. There is a second advantage to this mapping of the outcome into a number. In some experiments the investigator may not be interested in the outcome per se, but may be interested in some property associated with the outcome. It may be that the property or properties under study are common to some outcomes that are considered distinct in Ω. For example, if one is interested in the heights to the nearest inch of a group of people, all of the people (elements of Ω) are distinct, but for this study, those of the same height may be considered the same. Generally it is easy to define a function that maps distinct outcomes into the same number. With the above discussion in mind a random variable is defined as follows.

Definition I.1.4.   *A function that maps a sample space into the real numbers is called a* random variable.

We see that defining a random variable on a sample space may serve two purposes.

(i)   Each outcome is renamed to be some real number.

(ii)   Some of the useless information contained in the outcome is conveniently lost in the mapping from Ω into R.

Notation. $X(\omega)$, $Y(\omega)$, $Z(\omega), \ldots$ will be used to denote random variables. Generally the argument ω will be suppressed and we will represent the random variable by $X, Y, Z, \ldots$.

Example I.1.4.   Consider the experiment of tossing a coin ten times. The sample space for this experiment would consist of 2^{10} 10-tuples of heads and tails. That is $\Omega = \{(\omega_1, \omega_2, \ldots, \omega_{10}) : \omega_i \in \{H, T\}\}$. Now if one is interested in the number of heads obtained in the ten tosses, the number of distinct outcomes is no longer 2^{10} but rather only 11. Hence, if one is only interested in the number of heads obtained, it is natural to define for each $\omega \in \Omega$, $X(\omega)$ equals the number of heads obtained. This random variable serves both purposes listed above.

Example I.1.5. In some experiments there is not a natural number to assign to an outcome. For example, assume manufactured items are tested and rated as good, fair, and poor. In this case the sample space might be defined as $\Omega = \{G, F, P\}$. A typical random variable on this space would be $X(G) = 1$, $X(F) = 0$, and $X(P) = -1$. This random variable only serves purpose (i).

Example I.1.6. Consider the experiment of tossing a die once. In this case the sample space might be defined as $\Omega = \{1, 2, 3, 4, 5, 6\}$. If one is interested in whether or not the outcome is odd or even, a typical random variable for this problem would be $X(1) = X(3) = X(5) = 1$, $X(2) = X(4) = X(6) = 0$. Note that in this case the sample space was given to be a set of real numbers. Yet in view of the question being considered, a random variable is still used to serve purpose (ii).

Notice that a random variable should not be used to eliminate information until it is certain that the information will not be needed. For example, if in Example I.1.4 the question of how many heads appeared on the odd trials were asked, the random variable X would be worthless. In such a case one would be forced to return to the original sample space and define a second random variable for this second question. In practice, using two random variables to handle two separate problems may be efficient but the warning here is that in defining X one must be careful not to throw away information in ω that may be needed at another time.

By using a random variable, X, we change the focus of our study from the space Ω to the space S, where S is defined to be the range of X and hence is a subset of the real numbers. However, knowing that S contains the numbers 1, 12, and 30, say, is of questionable value unless the probability of each of these numbers occurring is also known. Since the original sample space, Ω, was endowed with a probability measure, there should be some mechanism for providing the space, S, with such a measure. If the probability measure on S is denoted by P^*, then the natural way to define a probability measure on $S = \{s_1, s_2, \ldots, s_n\}$ is to define $P^*(s_i) = P[\omega : X(\omega) = s_i]$. This measure on the sample space, S, is often called the *distribution of* X. In this way a new probability space, (S, P^*) can be constructed from (Ω, P) by using the random variable, X.

Example I.1.7. Consider Example I.1.1 again and assume you are interested in the number of heads that appear. A random variable that could be defined on Ω in this case is $X(\omega) = $ number of heads in ω. Hence the 16 points in Ω are mapped into the five points $\{0,1,2,3,4\}$. Now if the coin is balanced and each point in Ω has probability $1/16$, what is the probability measure on $S = \{0,1,2,3,4\}$? By using the binomial theorem or by simply counting we get

$$P^*(0) = 1/16$$

$$P^*(1) = 4/16$$

$$P^*(2) = 6/16$$

$$P^*(3) = 4/16$$

$$P^*(4) = 1/16.$$

Example I.1.8. Consider a classroom of 17 males and 13 females. Assume three students are to be chosen from this class without replacement and we are interested in the number of females chosen. The sample space for this experiment would be $\Omega = \{\omega : \omega = \text{a group of three students chosen from the } 30\}$. There are

$$\binom{30}{3} = \frac{30 \cdot 29 \cdot 28}{3 \cdot 2 \cdot 1} = 4060$$

points in Ω. Each of these points is assumed to be equally likely so they each have probability $1/4060$. If a random variable, X, is defined on Ω as $X(\omega) = $ number of females in the group, then the corresponding sample space is $S = \{0,1,2,3\}$. Using combinatorics, the probability measure P^* on S can be seen to be

$$P^*(0) = \frac{\binom{13}{0}\binom{17}{3}}{\binom{30}{3}} \approx .16$$

$$P^*(1) = \frac{\binom{13}{1}\binom{17}{2}}{\binom{30}{3}} \approx .43$$

$$P^*(2) = \frac{\binom{13}{2}\binom{17}{1}}{\binom{30}{3}} \approx .33$$

$$P^*(3) = \frac{\binom{13}{3}\binom{17}{0}}{\binom{30}{3}} \approx .07.$$

In most cases the analysis of an experiment with outcomes $\omega \in \Omega$ is done on the probability space (S, P^*) rather than on (Ω, P). This change of probability spaces from (Ω, P) to (S, P^*) is the reason for defining random variables. Unfortunately, in many instances the change is made so automatically that the reader only sees (S, P^*). The original sample space and associated probability are suppressed. The purpose of this section has been to show how and why this change is made. In subsequent chapters we too will often suppress the Ω space, but the reader should occasionally remind himself that the start of a probability problem is (Ω, P).

EXERCISES

1. Let $\Omega = \{a, b, c, d\}$ be a sample space. Determine whether or not each of the following functions defines a probability measure on Ω. If it does define a probability measure, find the probability of every point in Ω.

   i. $P_1(a) = \frac{1}{2}$, $P_1(a \cup c) = \frac{1}{3}$, $P_1(b) = \frac{1}{4}$

   ii. $P_2(b) = \frac{1}{3}$, $P_2(b \cup d) = \frac{1}{2}$, $P_2(a) = \frac{1}{2}$

   iii. $P_3(a \cup b \cup c) = \frac{1}{2}$, $P_3(a) = \frac{1}{6}$, $P_3(c) = \frac{1}{3}$

   iv. $P_4(a) = \frac{1}{2}$, $P_4(b) = \frac{1}{4}$, $P_4(c) = \frac{1}{3}$

   v. $P_5(a \cup b) = 0$, $P_5(c) = \frac{1}{4}$.

2. From a class of ten students, a president, vice president, secretary, and treasurer are selected. Define an appropriate sample space for this experiment.

3. Let $\Omega = \{\omega_1, \omega_2, \ldots, \omega_{10}\}$ be a class consisting of 3 freshmen, 2 sophomores, 1 junior, and 4 seniors. Let a random variable X on Ω be defined as $X(\omega_i) = $ year in school for the student, ω_i. If $P(\omega_i) = 1/10$ for $i = 1, 2, \ldots, 10$, describe the new probability space, (S, P^*) that one gets from X and (Ω, P).

4. An experimenter tosses a die three times. An appropriate sample space for this experiment would be $\Omega = \{(x_1, x_2, x_3) : x_i \in \{1, 2, 3, 4, 5, 6\}\}$. Define an appropriate random variable on Ω if the experimenter is interested in:

   i. the sum of the three tosses,
   ii. the average of the three tosses,
   iii. the largest of the three tosses.

SECTION 2   **STOCHASTIC PROCESSES**

Definition I.2.1. *A* stochastic process *is a family of random variables defined on some sample space, Ω. If there are countably many members of the* *family, the process will be denoted by* $X_1, X_2, X_3, \ldots$. *If there are uncountably many members of the family, the process will be denoted by* $\{X_t : t \geqslant 0\}$ *or* $\{X_t\}_{t \geqslant 0}$. *In the first case the process is called a* discrete-time *process, while in the second case it is called a* continuous-time *process.*

Definition I.2.2. *The set of distinct values assumed by a stochastic process is called the* state space. *If the state space of a stochastic process is countable, or finite, the process will be called a* chain. *The state space corresponds to the space S defined in the previous section.*

In this book most of the emphasis is placed on finite chains where the sample space and hence state space is finite. In cases where the sample space is uncountable, one needs an understanding of measure theory to study the process.

A stochastic process is sometimes viewed as a function of two variables, $X_t(\omega) = X(t, \omega)$. For fixed t the function is a random variable. For fixed ω, the resulting real-valued function of t is called a *sample path*.

Example I.2.1. Let $\{Y_i\}$, $i = 1, 2, \ldots, 6$ be a sequence of independent random variables with $P[Y_i = 1] = P[Y_i = -1] = \frac{1}{2}$ for $i = 1, 2, \ldots, 6$. Define $X_0 = 0$ and $X_n = \sum_{i=1}^{n} Y_i$. Now $X_n \equiv X_n(\omega) \equiv X(n, \omega)$ is a discrete-time stochastic process with $S = \{-6, -5, \ldots, -1, 0, 1, \ldots, 6\}$ as its state space. This stochastic process arises naturally from the following experiment. A fair coin is tossed six times and the player wins a dollar when a head appears and loses a dollar when a tail appears. Now X_n denotes the player's winnings at time n. A typical ω for this experiment is $\omega^* = (H, H, T, H, T, H)$. The corresponding sample path for this fixed ω^* is given in Figure I.2.1. The actual graph would consist of the vertices of the

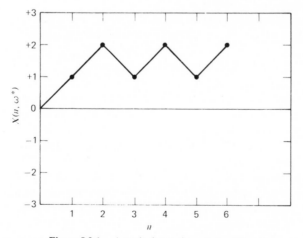

Figure I.2.1. A typical sample path for $X(n, \omega)$.

broken line but the lines are included for easier interpretation. There are 64 such paths corresponding to the 64 points in Ω.

Now consider the random variable $X_3(\omega)$. Recall that when the time variable is fixed at 3, the stochastic process becomes a random variable. The values assumed by $X_3(\omega)$ are $S^* = \{-3, -1, +1, +3\}$. The probability measure defined on this state space or equivalently the distribution of $X_3(\omega)$ is

$$P[X_3 = -3] = P^*(-3) = \binom{3}{0}(\tfrac{1}{2})^3 = \tfrac{1}{8},$$

$$P[X_3 = -1] = P^*(-1) = \binom{3}{1}(\tfrac{1}{2})^3 = \tfrac{3}{8},$$

$$P[X_3 = 1] = P^*(+1) = \binom{3}{2}(\tfrac{1}{2})^3 = \tfrac{3}{8},$$

$$P[X_3 = 3] = P^*(+3) = \binom{3}{3}(\tfrac{1}{2})^3 = \tfrac{1}{8}.$$

The random variables X_1, X_2, X_4, X_5, X_6 could also be considered and their distributions found (Exercise 6).

Example I.2.2. Assume the number of accidents occurring at a particular intersection is being observed for 30 days. Assume the waiting times between accidents are independent exponential random variables with mean $1/\lambda$ and let $N_t = N(t, \omega)$ denote the number of accidents that have occurred by time t. In this case N_t is called a Poisson process with parameter λ since for any fixed t^*, N_{t^*} has a Poisson distribution with mean λt^*. (See, for example, Parzen, 1962, p. 135.) The sample paths associated with this stochastic process would be nondecreasing with unit jumps randomly placed in $[0, 30]$. See Figure I.2.2 for a typical sample path for a Poisson process.

Note that the process defined in Example I.2.1 is a discrete-time stochastic process while the process in Example I.2.2 is a continuous-time stochastic process. Both of these processes have a countable state space.

In studying stochastic processes, one of the basic problems is that of finding the probability that the process is in some particular state at some fixed time. A variation of this problem is to find the probability that the process is in some particular state at a fixed time given some additional information as to the states the process has been in at some previous times. For example, in Example I.2.1 one might consider the probability that the process is in state 4 at time $n = 6$. A related problem would be to find the

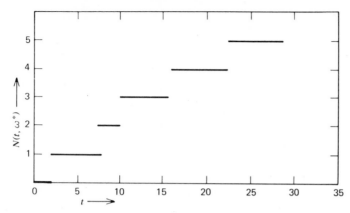

Figure I.2.2. A typical sample path for a Poisson process.

probability that the process is in state 4 at time $n = 6$ given that the process was in state 2 at time $n = 2$. The information about the process at time $n = 2$ does change the probability that the process will be in state 4 at time $n = 6$ and thus leads us to a consideration of the topic of conditional probability.

It is assumed that the reader has studied the concept of conditional probability previously, but since the concept is so central to this book, the basic definitions and ideas are reviewed here.

Let (Ω, P) be a probability space and let E and F be two subsets of Ω.

Definition I.2.3. *The symbol $P(E \cap F)$ denotes the probability that both E and F occur. A second notation for this probability is $P(E, F)$.*

Definition I.2.4. *If $P(F) \neq 0$, the probability of E given that F has occurred is defined to be*

$$P(E|F) = \frac{P(E \cap F)}{P(F)}. \tag{2.1}$$

The following example illustrates the use of (2.1) and may also give an intuitive feeling for conditional probability.

Example I.2.3. Toss two balanced dice and find the probability that the sum is at least 10 given that the first die is a 6. In this example the sample space Ω, for the experiment consists of 36 ordered pairs (x, y) where $x \in \{1, 2, 3, 4, 5, 6\}$ and $y \in \{1, 2, 3, 4, 5, 6\}$. Since the dice are balanced, the probability of each of these ordered pairs is $1/36$. A list of the sample points in Ω is given below:

$$
\begin{array}{cccccc}
(1,1) & (1,2) & (1,3) & (1,4) & (1,5) & (1,6) \\
(2,1) & (2,2) & (2,3) & (2,4) & (2,5) & (2,6) \\
(3,1) & (3,2) & (3,3) & (3,4) & (3,5) & (3,6) \\
(4,1) & (4,2) & (4,3) & (4,4) & (4,5) & (4,6) \\
(5,1) & (5,2) & (5,3) & (5,4) & (5,5) & (5,6) \\
(6,1) & (6,2) & (6,3) & (6,4) & (6,5) & (6,6)
\end{array}
\tag{2.2}
$$

The probability that the sum of two balanced dice is at least 10 is 6/36 since 6 of the 36 pairs have a sum of 10 or more. However, if it is given that the first die is a 6, the sample space under consideration becomes only the 6th row of (2.2). Looking at the 6th row we see that 3 of the 6 ordered pairs have a sum at least 10. Hence the probability of getting a sum of at least 10 given that the first die is a 6 is $\frac{3}{6} = \frac{1}{2}$. Now this technique of finding the new sample space under the side condition is fine for simple problems. However, in many cases this approach to conditional probability is difficult. That is, it may be difficult to define the new sample space and new probability measure under the side condition. Fortunately (2.1) yields a method of calculating the conditional probability using the original probability space. For this example

$$
P[\text{sum} \geqslant 10|\text{first die is } 6] = \frac{P[\text{sum} \geqslant 10 \text{ and first die} = 6]}{P[\text{first die} = 6]}
$$

$$
= \frac{3/36}{6/36} = \tfrac{1}{2}.
$$

Various exercises on conditional probability are given at the end of this section to assist the reader in reviewing this concept.

Equation 2.1 can also be used to find $P(E, F)$. This may be useful in cases where the events E and F have a natural ordering so it is easy to find $P(E)$ and $P(F|E)$ from which $P(E, F)$ can be calculated.

Example I.2.4. Consider again Example I.2.1. In order to find $P[X_6 = 4, X_2 = 2]$ simply calculate $P[X_2 = 2] = \frac{1}{4}$ and $P[X_6 = 4|X_2 = 2] = 3/16$ and then use (2.1).

Example I.2.5. Assume urn I contains 3 red and 2 green balls. Assume urn II contains 4 red and 8 green balls. If urn I is chosen with probability $\frac{1}{3}$ and the balls in each urn are chosen at random, find the probability of drawing a red ball. The event of drawing a red ball is split into the two mutually exclusive events, (urn I ∩ red ball) and (urn II ∩ red ball).

We have

$$P[\text{urn I, red ball}] = P[\text{urn I}] \cdot P[\text{red ball} \mid \text{urn I}]$$

$$= \tfrac{1}{3} \cdot \tfrac{3}{5} = \tfrac{1}{5}.$$

Similarly

$$P[\text{urn II, red ball}] = P[\text{urn II}] \cdot P[\text{red ball} \mid \text{urn II}]$$

$$= \tfrac{2}{3} \cdot \tfrac{1}{3} = \tfrac{2}{9}$$

and so

$$P[\text{red ball}] = \tfrac{1}{5} + \tfrac{2}{9} = 19/45.$$

In subsequent chapters, Equation 2.1 will be used most often as in Example I.2.5. That is, the joint probability of several events will be calculated using conditional probabilities. In fact, this will be done in the next section where the Markov property and transition matrices are defined.

EXERCISES

1. A certain item is manufactured at the rate of one per minute. Each item is inspected and rated good or defective. Describe the sample space for this experiment if the process runs for 10 minutes. A stochastic process is defined on this sample space as follows: $X_n(\omega)$ = number of defectives through time n. Draw a typical sample path for this process. If the probability of a defective item is $\tfrac{1}{3}$, find the distribution of the random variable $X_4(\omega)$.

2. If two cards are drawn without replacement from a standard bridge deck, find

   i. $P[\text{two aces} \mid \text{one card is the ace of spades}]$.

   ii. $P[\text{two aces} \mid \text{at least one of the cards is an ace}]$.

   (The fact that these questions have different answers shows that conditional probability can defy intuition.)

3. Consider an experiment that consists of tossing a balanced coin five times. Find

   i. $P[\text{4 heads} \mid \text{at least one head}]$.

   ii. $P[\text{at least one tail} \mid \text{at least one head}]$.

4. Referring to Example I.2.3, find the probability that the sum of X and Y exceeds 8 given that one or more of X and Y is a six.

5. Using the data of Example I.2.5, find P[urn I | red ball]. (Actually this problem is usually given as an application of Bayes' Theorem, but it can be done directly using the definition of conditional probability and the discussion in Example I.2.5.)

6. Consider the stochastic process described in Example I.2.1. Find the distribution of the random variable X_4. Show that $X_4 - X_2$ and $X_2 - X_0$ are independent and identically distributed.

SECTION 3   **THE MARKOV PROPERTY AND TRANSITION MATRICES**

Within the class of stochastic processes as defined earlier, there are many subclasses that have been studied in detail. In this section we define one of these subclasses by placing restrictions on the processes to be considered. The first restriction is that the process be a discrete-time process. The second is that only processes that have a countable or finite state space, will be considered. With these restrictions the notation for these processes will be $\{X_n(\omega)\}$ or $\{X_n\}$. Also, since the state space is a countable subset of the real numbers it is often convenient to let $S = \{1, 2, 3, \dots\}$. In any case an arbitrary state in S will be denoted by one of the lettters $\{i, j, k, l, m, n\}$. The final restriction that is imposed is that the process satisfy the Markov property. Before giving a definition of the Markov property, we will illustrate it by using a previous example.

Consider again Example I.2.1 where we now assume that the time parameter, n, is allowed to go beyond 6. X_n denotes the player's winnings at time n and let us say that $X_{n-1} = 10$. It is easy to see that X_n will be 9 or 11 since $P[Y_n = 1] = \frac{1}{2} = P[Y_n = -1]$. In fact, this relationship can be expressed in terms of conditional probability as

$$P[X_n = 9 | X_{n-1} = 10] = \tfrac{1}{2} \quad \text{and} \quad P[X_n = 11 | X_{n-1} = 10] = \tfrac{1}{2}.$$

Furthermore, the player's total winnings at time n depend only on X_{n-1}, the total winnings at time $n-1$, and the value of Y_n, the amount won at time n. The values of $X_{n-2}, X_{n-3}, \dots, X_1$ in no way affect the value of X_n. Expressing this in terms of conditional probability, we have

$$P[X_n = 9 | X_{n-1} = 10] = P[X_n = 9 | X_{n-1} = 10, X_{n-2} = i_{n-2}, \dots, X_1 = i_1].$$

This is an example of a stochastic process that satisfies the Markov property.

Definition I.3.1. *A stochastic process* $\{X_k\}$, $k = 1, 2, \ldots$ *with state space* $S = \{1, 2, 3, \ldots\}$ *is said to satisfy the Markov property if for every n and all states $i_1, i_2, \ldots, i_n$ it is true that*

$$P[X_n = i_n | X_{n-1} = i_{n-1}, X_{n-2} = i_{n-2}, \ldots, X_1 = i_1] = P[X_n = i_n | X_{n-1} = i_{n-1}].$$

$$(3.1)$$

As a further illustration, consider a mouse moving through the maze shown in Figure I.3.1 (see Karlin, 1968). Assume that the mouse moves from room to room by choosing at random one of the doors available to him. (We are assuming that this is an honest mouse and that he will in fact choose the doors with equal probability!) Finally, assume that the mouse changes rooms at specified times $n = 1, 2, 3, \ldots$. Now define $X_n =$ number of the room occupied by the mouse at time n. The state space S for this problem is $S = \{1, 2, \ldots, 9\}$. From the assumptions made in this example, it is clear that the probability that the mouse goes to room i_n at time n depends only on his location at time $n - 1$ and not on his location at earlier times. That is, $\{X_k\}$ satisfies the Markov property.

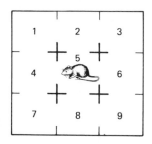

Figure I.3.1. A mouse in a maze.

We will not always speak in terms of a mouse going through a maze, but we will often speak of a "particle" moving among the states of S. (This kind of terminology may not be precise mathematically, but it is helpful in getting a mental picture of certain behavior.) If we say "the particle is in state i at time $n - 1$," we simply mean $X_{n-1} = i$.

If we know that the particle is in state i at time $n - 1$, we might ask where it will be at time n. Generally a precise answer to this question can not be given; rather we must give a probability distribution over the states in S, conditional on the particle's location at time $n - 1$. This conditional distribution is given by

$$P[X_n = j | X_{n-1} = i].$$

We might also ask whether or not knowledge of where the particle was *before* time $n-1$ would alter this conditional probability. If such knowledge does not alter the conditional probability distribution, the Markov property is satisfied. Roughly speaking, the Markov property is satisfied if the future location of the particle depends on its present location, but not its past locations.

A stochastic process that satisfies the three restrictions given at the beginning of this section is called a discrete-time Markov chain. The phrase discrete time comes from restriction one, the word chain comes from restriction two, and the word Markov comes from restriction three.

Example I.3.1. Consider a basketball game for which the following stochastic process is defined. $X_n(\omega) = $ the score of team A minus the score of team B at the time of the nth score. That is, the process will be viewed as a discrete-time process by ignoring elapsed time between scores. Using this definition of time, the process is clearly a discrete-time chain.

The question of whether or not it is Markov is highly debatable for several reasons.

To see if it is reasonable to assume that the Markov property holds, consider the events $[X_n = 2 | X_{n-1} = 0]$, $[X_n = 2 | X_{n-1} = 0, X_{n-2} = 2]$, and $[X_n = 2 | X_{n-1} = 0, X_{n-2} = -2]$. In the context of this example the events are "team A scores given that the score is tied," "team A scores given that the score is tied and team B just tied the score," and "team A scores given that the score is tied and team A just tied the score." Now if there is indeed an advantage to having the ball, then since the ball changes hands after a score one might reasonably assign different probabilities to these three events, in which case the Markov property would fail to hold.

To carry this same theme a bit further, if a person believed that "momentum" actually plays a role in the game of basketball, then it might be that values even further in the past could affect the probabilities assigned to events. For example the events $[X_n = 2 | X_{n-1} = 0, X_{n-2} = 2, X_{n-3} = 4, X_{n-4} = 6]$ and $[X_n = 2 | X_{n-1} = 0, X_{n-2} = -2, X_{n-3} = -4, X_{n-4} = -6]$ might be assigned different probabilities than the latter two events defined in the previous paragraph. Hence we see that considering ball possession and momentum, some people might be willing to consider $\{X_n\}$ a Markov chain while many others would not. This example points out the basic problem faced by anyone who wishes to use the theory of Markov chains to study a stochastic process. It is seldom completely clear whether or not the Markov property is in fact satisfied. This is not a problem that will be emphasized in this book but it is mentioned as the crucial step in trying to apply the theory in this book to some experiment. There is extensive literature on this general problem of testing whether or not the Markov assumption is valid for a process.

Once a stochastic process falls into the subclass of a discrete-time Markov chain, the movement of the process among the states of S is determined by the conditional probabilities, $P[X_n = j | X_{n-1} = i]$. For notational convenience these probabilities are often written as $p_{ij}^{(n-1,n)}$. The next question we ask is whether these conditional probabilities do depend on the time at which the transition from i to j takes place. This leads to the following definition.

Definition I.3.2. *A discrete-time Markov chain is said to be* stationary *or* homogeneous in time *if the probability of going from one state to another is independent of the time at which the step is being made. That is, for all states i and j,*

$$P[X_n = j | X_{n-1} = i] = P[X_{n+k} = j | X_{n+k-1} = i] \tag{3.2}$$

for $k = -(n-1), -(n-2), \ldots, -1, 0, 1, 2, \ldots$. The Markov chain is said to be nonstationary *if the condition for stationarity fails.*

Example I.3.2. Assume a machine is producing items independently at the rate of one a minute. Let X_n denote the number of defectives produced by time n. If the probability of producing a defective item remains constant throughout the life of the machine, then X_n would be a stationary Markov chain. However, if the probability of producing a defective item changes as the machine grows older, then the Markov chain would be nonstationary.

Example I.3.3. Consider a gambler playing the game described in Example I.2.1 with $X_0 = 0$ and $X_n =$ winnings at time n. Since the probability of winning does not change with time, this Markov chain is stationary.

In the case of a stationary Markov chain, the notation introduced above to represent transition probabilities can be further simplified. We defined $p_{ij}^{(n-1,n)}$ to mean the probability of going from i to j on the nth step. For a stationary chain the dependence on n can be dropped and p_{ij} can be used to denote the probability of going from i to j. Recall that p_{ij} is actually a conditional probability with the following meaning: $p_{ij} = P$ [the process is in state i and goes to state j in the next step]$/P$ [the process is in state i]. These conditional probabilities are often called *transition probabilities* for the chain.

Let $\{X_k\}$ denote a discrete-time stationary Markov chain with a finite state space, $S = \{1, 2, \ldots, n\}$. For this chain there are n^2 transition probabilities, $\{p_{ij}\}$ $i = 1, 2, \ldots, n; j = 1, 2, \ldots, n$. The most convenient way of recording these values is in the form of a matrix, P. Associate the ith row and column of P with the ith state of S and the matrix takes the form

$$P = \begin{bmatrix} p_{11} & p_{12} & \cdots & p_{1n} \\ p_{21} & p_{22} & \cdots & p_{2n} \\ \vdots & & & \vdots \\ p_{n1} & p_{n2} & \cdots & p_{nn} \end{bmatrix}.$$

This matrix is called the *transition probability matrix* or the *transition matrix* corresponding to the discrete-time stationary Markov chain $\{X_k\}$. This matrix contains all the relevant information regarding the movement of the process among the states in S. In fact our study of discrete-time stationary Markov chains will soon be reduced to a study of the corresponding transition matrix, P.

Note that every transition matrix has the following properties:

(i) all the entries are non-negative,
(ii) the sum of the entries in each row is one.

The first property is obvious and the second can be shown as follows:

$$p_{i1} + p_{i2} + \cdots + p_{in}$$

$$= P[X_k = 1 | X_{k-1} = i] + P[X_k = 2 | X_{k-1} = i] + \cdots + P[X_k = n | X_{k-1} = i]$$

$$= P[(X_k = 1) \cup (X_k = 2) \cup \cdots \cup (X_k = n) | X_{k-1} = i]$$

$$= P[X_k \in S | X_{k-1} = i] = 1.$$

Definition I.3.3. *Any square matrix that satisfies conditions* (i) *and* (ii) *above is called a* stochastic matrix.

There is a subtle distinction between a transition matrix and a stochastic matrix. When referring to a transition matrix, one usually has a specific Markov chain in mind. That is, a definite sequence of random variables and corresponding sample space are being considered. On the other hand, the phrase stochastic matrix is used to refer to a matrix with properties (i) and (ii) where no attempt is made to relate this to any specific Markov chain. Results that are proved for stochastic matrices may be directly applied to Markov chains by simply noting that a transition matrix is a stochastic matrix. Hence when developing the general theory of Markov chains, stochastic matrices will be used since one does not want the results to depend on any particular process. However, this does not mean that specific transition matrices are unimportant. As mentioned previously, the first question faced by someone using the theory of Markov chains is whether or not the process is Markov. If he believes he is working with a discrete-time stationary Markov chain, the next step is to find the transition matrix. The following examples show how this can be done.

Example I.3.4. Consider two urns containing b balls each. Assume a ball is chosen at random with probability $1/2b$. (This can be done by numbering the balls initially and choosing a number randomly.) Take the selected ball and place it into the other urn. Let X_n denote the number of balls in urn I at time n. This yields a discrete-time stationary Markov chain with transition matrix given by

$$
P = \begin{pmatrix}
0 & 1 & 0 & 0 & \cdots & \cdots & \cdots & 0 & 0 & 0 \\
\frac{1}{2b} & 0 & 1-\frac{1}{2b} & 0 & \cdots & \cdots & \cdots & 0 & 0 & 0 \\
0 & \frac{2}{2b} & 0 & 1-\frac{2}{2b} & \cdots & \cdots & \cdots & 0 & 0 & 0 \\
\cdots & \cdots & \cdots & \cdots & \cdots & \cdots & \cdots & \cdots & \cdots & \cdots \\
0 & \cdots & \cdots & \cdots & \cdots & \cdots & \cdots & 1-\frac{1}{2b} & 0 & \frac{1}{2b} \\
0 & \cdots & \cdots & \cdots & \cdots & \cdots & \cdots & 0 & 1 & 0
\end{pmatrix}.
$$

There are $2b+1$ rows and columns in P where the ith row and column correspond to $i-1$ balls in urn I. This particular Markov chain is referred to as the Ehrenfest diffusion model. [See Feller (1968) for a more complete discussion of this problem.]

Example I.3.5. Assume a fair coin is tossed repeatedly and X_n denotes the number of heads obtained by the nth toss. This process defines a discrete-time stationary Markov chain. If there is no bound on the number of tosses, the corresponding transition matrix will be infinite. In particular P is given by

$$
P = \begin{pmatrix}
\frac{1}{2} & \frac{1}{2} & 0 & 0 & \cdots \\
0 & \frac{1}{2} & \frac{1}{2} & 0 & \cdots \\
0 & 0 & \frac{1}{2} & \frac{1}{2} & \cdots \\
\cdots & \cdots & \cdots & \cdots & \cdots \\
\cdots & \cdots & \cdots & \cdots & \cdots
\end{pmatrix}.
$$

Again the ith row and column of P correspond to $i-1$ heads.

Example I.3.6. Toss a red die and a green die together. Let the outcome of the ith toss be recorded as the ordered pair, (R_i, G_i). Note that $R_i \in \{1,2,3,4,5,6\}$ and $G_i \in \{1,2,3,4,5,6\}$. Define $Z_n = \sum_{i=1}^{n}|R_i - G_i|$ and $Z_0=0$. Then $\{Z_n\}$ is a discrete-time stationary Markov chain and if both dice are assumed to be balanced, the transition matrix can be found using elementary probability theory (Exercise 2).

In the three previous examples it was not too hard to determine the appropriate state space and the transition probabilities necessary for describing the Markov chain of interest. However, in some situations this determination is quite difficult. In fact in the following example it is not even apparent that the problem can be solved by using Markov chain techniques!

Example I.3.7. Let a fair die be tossed repeatedly with X_i defined to be the number appearing up on the ith toss. Define S_n to be the sum of the first n X_i's. Let k be fixed and define α_k by

$$\alpha_k = P[S_n = k \text{ for some } n].$$

We wish to find $\lim_{k \to \infty} \alpha_k$.

In looking for a Markov chain to associate with this problem it is natural to try to use the variables S_n. These do indeed form a Markov chain with state space $\{1, 2, 3, \ldots\}$ and transition probabilities $p_{ij} = \frac{1}{6}$ if $1 \leqslant j - i \leqslant 6$. However, this chain does not help in finding the desired limit. The following solution to the problem using a stochastic matrix corresponding to a Markov chain with six states was shown to us by Professor H. T. David.

For a given k, let n_k be the first number such that S_n is greater than or equal to k, that is

$$n_k = \min\{n : S_n \geqslant k\}.$$

Define $D_k = S_{n_k} - k$, so that D_k is the excess over k after n_k trials. Clearly $D_k \in \{0, 1, \ldots, 5\}$. If we define

$$p_k(j) = P[D_k = j], \qquad j = 0, 1, \ldots, 5,$$

then we have that $p_k(0) = \alpha_k$.

If $D_k = j + 1$, then it must be that $D_{k+1} = j$ for $j = 0, 1, \ldots, 4$ and if $D_k = 0$, then D_{k+1} is equally likely to be 0, 1, 2, 3, 4, or 5. In terms of conditional probability this says

$$P[D_{k+1} = j | D_k = j + 1] = 1, \qquad j = 0, 1, \ldots, 4$$

and

$$P[D_{k+1} = j | D_k = 0] = \tfrac{1}{6}, \qquad j = 0, 1, \ldots, 5.$$

If we treat the values of D_k as the states of a space, namely $S = \{0, 1, \ldots, 5\}$, then $\{D_k\}$ forms a Markov chain with transition matrix

$$P = \begin{pmatrix} \frac{1}{6} & \frac{1}{6} & \frac{1}{6} & \frac{1}{6} & \frac{1}{6} & \frac{1}{6} \\ 1 & 0 & 0 & 0 & 0 & 0 \\ 0 & 1 & 0 & 0 & 0 & 0 \\ 0 & 0 & 1 & 0 & 0 & 0 \\ 0 & 0 & 0 & 1 & 0 & 0 \\ 0 & 0 & 0 & 0 & 1 & 0 \end{pmatrix}.$$

To complete the solution to this problem requires the use of techniques which will be developed later in the book. In Exercise III.2.15 the reader will be asked to use P to find $\lim_{k \to \infty} p_k(0) = \lim_{k \to \infty} \alpha_k$.

In Examples I.3.4–I.3.7 the appropriate transition matrices can be determined by using elementary probability theory. In a sense, the entries in the transition matrices are provided for us by Mother Nature. Unfortunately, this is not always the case. The following example illustrates a situation where probability theory, elementary or otherwise, cannot be used to determine the transition matrix. This example also illustrates a method that may be used to determine the transition matrix.

Example I.3.8. Blumen, Kogan, and McCarthy (1955) considered the question of the flow of males aged 20–24 through various occupations. They first classified occupations into 11 categories. These categories represent the 11 states of the chain. (In our notation these states would be called $\{1, 2, 3, \ldots, 11\}$.) The time unit chosen was three months since quarterly reports are available. Given that one agrees with the assumption that males will move among these states according to a discrete-time stationary Markov chain, how would one determine P? In this case, no general probability model is given from which p_{ij} can be exactly determined. The standard approach in such cases is to look at data from the past and use this to estimate p_{ij}. In particular p_{ij} is estimated by the ratio N_{ij}/N_i where N_{ij} is the total number of workers who moved during one quarter from state i to state j; N_i is the total number of workers who started the quarter in state i. It is a very important question whether or not these values are accurate in any sense. This question will not be answered here since it is statistical in nature and would require another book. However, the question is raised so the distinction between the theory and applications of Markov chains can be better understood. In studying the theory of Markov chains we simply *assume* the existence of the appropriate transition matrix, while in applications an appropriate transition matrix for the experiment in question must be found.

The transition matrix contains all the information needed to describe the motion of the chain among the states in S. However, if one is interested in where the process is at any particular time, he must first know where the chain started.

Definition I.3.4. *A vector* $\mathbf{a}_0 = (\alpha_1, \alpha_2, \ldots, \alpha_n)$ *is called a* starting vector *if* $\sum_{i=1}^{n} \alpha_i = 1$ *and* $\alpha_i \geq 0$ *for* $i = 1, 2, \ldots, n$.

In the case where the chain starts deterministically at one state, $\mathbf{a}_0$ has a one in the coordinate corresponding to that state and zeros elsewhere. In general, the process can start at various states according to some probability distribution. The starting vector gives this distribution. For notational convenience the starting vector will be referred to as the distribution at time zero. That is, $\alpha_k = P[X_0 = k]$, $k = 1, 2, \ldots, n$.

We now show how the starting vector and transition matrix can be used to determine the probabilities of various outcomes for the process. For example, if you want to find $P[X_0 = i_0, X_1 = i_1, \ldots, X_m = i_m]$, where $i_k \in \{1, 2, \ldots, n\}$ for $k = 1, 2, \ldots, m$, use the theory of conditional probability to simplify this as follows:

$$P[X_0 = i_0, X_1 = i_1, \ldots, X_m = i_m]$$
$$= P[X_0 = i_0]P[X_1 = i_1 | X_0 = i_0]P[X_2 = i_2 | X_0 = i_0, X_1 = i_1] \cdots$$
$$P[X_m = i_m | X_0 = i_0, X_1 = i_1, \ldots, X_{m-1} = i_{m-1}].$$

Since $\{X_k\}$ is assumed to be a discrete-time stationary Markov chain, this reduces to $\alpha_{i_0} p_{i_0 i_1} p_{i_1 i_2} \cdots p_{i_{m-1} i_m}$.

In many instances one is not interested in a step-by-step analysis of the chain but only in where the chain is after m steps. For example, consider the simple problem of finding $P[X_1 = i]$. Again using conditional probabilities this can be written as

$$P[X_1 = i] = P[X_0 = 1]P[X_1 = i | X_0 = 1] + P[X_0 = 2]P[X_1 = i | X_0 = 2]$$
$$+ \cdots + P[X_0 = n]\, P[X_1 = i | X_0 = n]$$
$$= \sum_{j=1}^{n} \alpha_j p_{ji}.$$

Similarly

$$P[X_2 = i] = \sum_{k=1}^{n} \sum_{j=1}^{n} \alpha_j p_{jk} p_{ki}.$$

Now it is very fortunate that formulas of this type need not be memorized.

It happens that the matrix notation introduced above for the transition matrix, P, is ideally suited to handling this problem. The expression for $P[X_1 = i]$ is simply the ith coordinate of the vector $\mathbf{a}_0 P$. Call this vector $\mathbf{a}_1$. This vector represents the distribution of where the Markov chain is after one step. As the reader would guess now, $\mathbf{a}_2 = (\mathbf{a}_0 P)P = \mathbf{a}_0 P^2$ where the ith coordinate of $\mathbf{a}_2$ is $P[X_2 = i]$. In general, if one wants the distribution of where the process is after n steps given that the starting vector was $\mathbf{a}_0$, simply find $\mathbf{a}_n = \mathbf{a}_0 P^n$. The matrix P^n that appears in this expression also has a special meaning. It can be shown that P^n is a transition matrix that represents the n step transition probabilities. (Exercise 7.) In view of this the elements of P^n will be denoted by $p_{ij}^{(n)}$ where $p_{ij}^{(n)} = P[X_{k+n} = j \mid X_k = i]$.

Example I.3.9. Consider Example I.3.4 with $b = 2$. In this case the transition matrices P and P^2 are given below where the ith row and column of P correspond to $i - 1$ balls in urn I:

$$
P = \begin{bmatrix} 0 & 1 & 0 & 0 & 0 \\ \frac{1}{4} & 0 & \frac{3}{4} & 0 & 0 \\ 0 & \frac{1}{2} & 0 & \frac{1}{2} & 0 \\ 0 & 0 & \frac{3}{4} & 0 & \frac{1}{4} \\ 0 & 0 & 0 & 1 & 0 \end{bmatrix}, \qquad P^2 = \begin{bmatrix} \frac{1}{4} & 0 & \frac{3}{4} & 0 & 0 \\ 0 & \frac{5}{8} & 0 & \frac{3}{8} & 0 \\ \frac{1}{8} & 0 & \frac{3}{4} & 0 & \frac{1}{8} \\ 0 & \frac{3}{8} & 0 & \frac{5}{8} & 0 \\ 0 & 0 & \frac{3}{4} & 0 & \frac{1}{4} \end{bmatrix}
$$

As the example is stated, the chain starts with 2 balls in each cell so $\mathbf{a}_0 = (0, 0, 1, 0, 0)$. Hence $\mathbf{a}_1 = (0, \frac{1}{2}, 0, \frac{1}{2}, 0)$ and $\mathbf{a}_2 = (\frac{1}{8}, 0, \frac{3}{4}, 0, \frac{1}{8})$. Admittedly it would be tedious to find $\mathbf{a}_{20}$ by hand, but this could easily be done on a computer. In any event, the notation of starting vectors and transition matrices greatly simplifies the general problem of deciding where a discrete-time stationary Markov chain will be after M steps.

Now there is one potential problem created by the introduction of starting vectors. For a stationary Markov chain we have that the probability of going from i to j does not depend on when the step is made. However, for the above example consider $P[X_2 = 1 \mid X_1 = 0]$. If one allows the starting vector to enter into the problem, he must say this conditional probability is undefined since $P[X_1 = 0 \mid X_0 = 2] = 0$. In view of this dilemma, one might be tempted to say the chain is in fact not stationary since it appears that $P[X_2 = 1 \mid X_1 = 0] \neq P[X_3 = 1 \mid X_2 = 0]$ when $\mathbf{a}_0 = (0, 0, 1, 0, 0)$. This is the wrong way to view the problem, however, and the warning here is that the starting vector should *not* be used in determining stationarity for a discrete-time stationary Markov chain. For the same reason the starting vector should not be used in the calculation of the transition probabilities.

This section is concluded by giving the Chapman—Kolmogorov identity for transition probabilities. In order to have a general statement of this identity we need to define $p_{ij}^{(0)}$.

Definition I.3.5. $p_{ij}^{(0)} = \delta_{ij}$ where $\delta_{ij} = 1$ if $i = j$ and $\delta_{ij} = 0$ if $i \neq j$.

Theorem I.3.1   (Chapman–Kolmogorov Identity).

For all non-negative integers m and l

$$p_{ij}^{(l+m)} = \sum_{k \in S} p_{ik}^{(l)} p_{kj}^{(m)}. \tag{3.3}$$

Proof.   A formal proof follows easily using mathematical induction (Exercise 8). A heuristic argument is given here since this argument is helpful in remembering the identity. The left-hand side represents the probability of going from i to j in $l + m$ steps. This amounts to measuring the probability of all those sample paths (ω's) that start at i and end at j after $l + m$ steps. The right-hand side of the Chapman–Kolomogorov identity takes this collection of paths and partitions it according to where the path was after l steps. All those paths that go from i to k in l steps and then from k to j in m steps are grouped together and the probability of this group of paths is given by $p_{ik}^{(l)} p_{kj}^{(m)}$. By summing these probabilities over all $k \in S$ we get the probability of going from i to j in $l + m$ steps. (The probabilities are summed since the groups associated with different k's are disjoint.) That is, in going from i to j in $l + m$ steps, the chain must be someplace in S after l steps. The right-hand side of (3.3) considers all the places it might be and uses that as a criterion for partitioning the set of paths that go from i to j in $l + m$ steps.   ▲

SUMMARY

The purpose of these last two sections has been to first give the reader some feeling for the structure of stochastic processes, including the relationship between the sample space, Ω, and state space, S. Once the class of stochastic processes was defined, our attention was restricted to smaller and smaller subclasses by defining discrete-time, stationary Markov chains. The reduction resulting from each of these words is probably best seen in Figure I.3.2.

The class of all Markov processes is given by

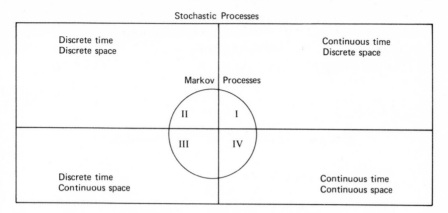

Figure I.3.2. Classes of stochastic processes.

The class of discrete-time Markov chains is given by the second quadrant, II.

The class of discrete-time stationary Markov chains is a subset of II.

For notational convenience the phrase "Markov chain" will be used in this book to denote a discrete-time stationary Markov chain. If a Markov chain is nonstationary or defined on continuous time, these properties will be explicitly stated.

We conclude Chapter I with a discussion of some results from mathematical analysis that are often not included in a basic course on advanced calculus. In any event the results discussed in the next section are used throughout the remainder of the book and the reader should be familiar with these results before proceeding to Chapter II.

EXERCISES

1. Find the transition matrix for the "mouse in the maze" problem.

2. Find the transition matrix for Example I.3.6.

3. There are 6 white marbles in urn A and 4 red marbles in urn B. At each step a marble is selected at random from each urn and the two marbles interchanged. Let X_n be the number of red marbles in urn A after n interchanges.

   i.   Does X_n form a stationary Markov chain?
   ii. If so, find the transition matrix P.
   iii. What is the probability that there are 2 red marbles in A after 3 steps?

4. A certain gambler does not believe in the independence of consecu-

tive tosses of a coin. Instead he strongly believes an honest coin should yield approximately half heads in the sense that if the number of tails exceeds the number of heads, the probability of heads on the next toss is greater than $\frac{1}{2}$. In fact he believes

$$P[\text{heads}|m \text{ heads and } k \text{ tails}] = \frac{k}{m+k} \qquad \text{if } m \neq 0 \text{ and } k \neq 0$$

$$P[\text{heads}|m \text{ heads and } k \text{ tails}] = \frac{k}{k+1} \qquad k \neq 0, m = 0$$

$$P[\text{heads}|m \text{ heads and } k \text{ tails}] = \frac{1}{m+1} \qquad m \neq 0, k = 0$$

$$P[\text{heads}|m \text{ heads and } k \text{ tails}] = \frac{1}{2} \qquad m = 0, k = 0$$

Let X_n be the stochastic process that denotes the number of heads in n tosses. Answer the following questions from the point of view of the gambler.

   i.   Is X_n a Markov chain?
   ii.   Find $P[2 \text{ heads after 3 tosses}]$ and $P[\text{no heads after 4 tosses}]$.
   iii.   If X_n is a Markov chain, is it stationary?
   iv.   If it is stationary, find the transition matrix, P.

5.   Let J consist of the $n!$ permutations of the integers $1, 2, \ldots, n$, and let the probability of a particular permutation $\omega = (a_1, a_2, \ldots, a_n)$ be $1/n!$. Random variables $\{X_n\}$ are defined as follows: $X_0(\omega) = 1$; $X_{k+1}(\omega)$ is the least $i > X_k(\omega)$ such that $a_i > a_{X_k(\omega)}$ if there is such an i, $1 \leqslant i \leqslant n$, otherwise $X_{k+1}(\omega) = n + 1$. $\{X_k\}$ is a stationary Markov chain with state space $1, 2, \ldots, n+1$. The problem is to find the transition probability matrix $P = \{p_{ij}\}$.

6.   Assume you want to study the movement of persons among various income groups from year to year. In particular take a state space consisting of the income groups

State	Income
1	0– 5,000
2	5,001–10,000
3	10,001–15,000
4	15,001–20,000
5	20,001–25,000
6	25,001–30,000
7	more than 30,000

(Assume all incomes are rounded up to the nearest dollar so all incomes fall into exactly one group.) Discuss the problems caused by inflation. That is if one wanted to view this model as a Markov chain, what problems would inflation cause and how could these problems be handled?

7. (a) Show that P^n is a stochastic matrix whenever P is.
   (b) Show the (i,j) entry of P^n is equal to $p_{ij}^{(n)}$.

*8. Prove the Chapman–Kolmogorov identity.

9. A generalization of the Markov property as defined in Definition I.3.1 is as follows:

A stochastic process $\{X_n\}, n = 1, 2, \ldots$ with state space $S = \{1, 2, 3, \ldots\}$ is said to be a kth-*order Markov chain* if for every n and all states $i_1, i_2, \ldots, i_n$ it is true that

$$P[X_n = i_n | X_{n-1} = i_{n-1}, X_{n-2} = i_{n-2}, \ldots X_{n-k} = i_{n-k} \ldots X_1 = i_1]$$

$$= P[X_n = i_n | X_{n-1} = i_{n-1}, X_{n-2} = i_{n-2}, \ldots X_{n-k} = i_{n-k}].$$

(a) Show that a kth-order Markov chain is also a $(k+1)$st-order chain for $k = 1, 2, \ldots$.
(b) Discuss whether or not it is reasonable to view the stochastic process defined in Example I.3.1 as a second-order Markov chain.

SECTION 4   **MATHEMATICAL ANALYSIS REVIEW**

In this section we will leave the topic of Markov chains per se in order to establish some mathematical results that will be needed throughout the book. Since we will concentrate on developing only those results that will be needed, this section may suffer from a lack of continuity. However it is expected that those who read this section have some background in advanced calculus so they can fill in the gaps.

The first and probably most important topic we consider is the definition of a limit. In particular, what is meant by saying that a sequence of numbers, $\{x_n\}$, converges to x_0 as n goes to infinity?

Definition I.4.1. *Let* $\{x_n\}$ *be a sequence of real numbers. Then* $\lim_{n\to\infty} x_n = x_0$ *if for every* $\epsilon > 0$, *there exists an integer* $N = N(\epsilon)$ *such that* $|x_n - x_0| < \epsilon$ *whenever* $n \geqslant N$.

Definition I.4.1′. *Let* $f(s)$ *be a real-valued function of* s *for* $s \in [a,b]$. *We say* $\lim_{s\to s_0} f(s) = t_0$ *if for every* $\epsilon > 0$, *there exists a* $\delta > 0$ *such that* $|f(s) - t_0| < \epsilon$ *whenever* $|s - s_0| < \delta$.

We are aware of the fact that these definitions are difficult for many students of advanced calculus and almost all students of calculus. In view of this and the fact that we will use these definitions repeatedly in this book, some discussion of Definition I.4.1 will be given here. For this discussion we will refer to the index n as a time parameter since this is consistent with our applications of this definition. One common way of describing Definition I.4.1 is to say that as the time parameter, n, gets larger and larger, the value of the sequence, x_n, gets closer and closer to x_0. The main fault with this interpretation of the definition is that it implies that x_{n+1} is closer to x_0 than x_n is (that is, $|x_{n+1} - x_0| < |x_n - x_0|$). This certainly need not be the case as is shown by the following example.

Example I.4.1. Let $x_{2n} = 0$ and $x_{2n-1} = 1/(2n-1)$ for $n = 1, 2, 3, \ldots$. In this case $\lim_{n \to \infty} x_n = 0$ but $|x_{2n} - 0| < |x_{2n+1} - 0|$.

In view of this example, we give a second phrasing of Definition I.4.1 that does not have this implication of monotonicity. The definition essentially says if someone tells you how close he wants the x_n sequence to get to x_0, you can give him a time such that beyond that time, the sequence remains at least that close to x_0. The measure of closeness is given by the small positive number, ϵ. The time beyond which this degree of closeness is guaranteed is given by $N = N(\epsilon)$. Note that the time beyond which x_n stays within ϵ of x_0 does depend on ϵ. (This is the reason for writing N as a function of ϵ). Now if this $N = N(\epsilon)$ can be found for every $\epsilon > 0$, then $\lim_{n \to \infty} x_n = x_0$.

Example I.4.2. Show that if $x_n = 1/n$, then $\lim_{n \to \infty} x_n = 0$. Let $\epsilon > 0$ be given. Now $1/n < \epsilon$ whenever $n > 1/\epsilon$. Hence choose $N(\epsilon)$ to be any integer greater than $1/\epsilon$. In this case $|x_n - 0| < \epsilon$ for all $n \geq N(\epsilon)$ so $\lim_{n \to \infty} x_n = 0$.

One method of showing that x_n converges to something is to show that the $\limsup_{n \to \infty} x_n$ and the $\liminf_{n \to \infty} x_n$ are equal. In order to understand this method, a short discussion of the terms $\limsup$ and $\liminf$ will be given. (The notations $\limsup_{n \to \infty} x_n = \overline{\lim}_{n \to \infty} x_n$ and $\liminf_{n \to \infty} x_n = \underline{\lim}_{n \to \infty} x_n$ are also used in this book.)

Example I.4.3. Consider the sequence $x_{2n} = 1$ and $x_{2n-1} = 1/(2n-1)$ for $n = 1, 2, 3, \ldots$. It is easy to see that this sequence does not converge. However, the values zero and one are in some sense related to the values of the sequence after a "long time." In particular $\lim_{n \to \infty} x_{2n} = 1$ and $\lim_{n \to \infty} x_{2n-1} = 0$. Since there are these subsequences of $\{x_n\}$ that converge to zero and one, we say that zero and one are *cluster points* of $\{x_n\}$. Intuitively speaking, the $\limsup_{n\to\infty} x_n$ is the largest of the cluster points of $\{x_n\}$. The $\liminf_{n\to\infty} x_n$ is the smallest of the cluster points of $\{x_n\}$. When the smallest cluster point is equal to the largest cluster point, this unique cluster point is the limit of the sequence. A rigorous definition of $\limsup_{n\to\infty} x_n$ and $\liminf_{n\to\infty} x_n$ depends on the concepts of supremum and infimum. The reader is referred to any textbook on advanced calculus for these definitions.

A similar discussion could be given for Definition I.4.1′ but since an understanding of Definition I.4.1′ usually follows easily from an understanding of I.4.1, only an example of I.4.1′ will be given.

Example I.4.4. Let $f(s) = \frac{3}{4} - s^2$ for $s \in [0, 1]$. Then $\lim_{s\to 1/2} f(s) = \frac{3}{4} - (\frac{1}{2})^2$ $= \frac{1}{2}$. To show that this is true let $\epsilon > 0$ be given. Consider

$$\left|\left(\tfrac{3}{4} - s^2\right) - \tfrac{1}{2}\right| = \left|\tfrac{1}{4} - s^2\right| = \left|\tfrac{1}{2} - s\right| \cdot \left|\tfrac{1}{2} + s\right| < \left|\tfrac{1}{2} - s\right| \cdot \left(\tfrac{3}{2}\right).$$

Now if $\left|\tfrac{1}{2} - s\right| < \epsilon \cdot \tfrac{2}{3}$, then $\left|f(s) - \tfrac{1}{2}\right| < \epsilon$. Hence by choosing $\delta = 2\epsilon/3$ we have the desired result.

Closely related to the concept of a sequence of real numbers is the concept of a series.

Definition I.4.2. *Let* $x_1, x_2, x_3, \ldots$ *be a sequence of real numbers. The series or* infinite sum *determined by this sequence is defined to be* $\sum_{i=1}^{\infty} x_i = x_1 + x_2 + x_3 + \cdots$.

The first question asked about a series is whether or not it converges. In order to define what this means we consider the sequence of partial sums determined by the sequence $\{x_n\}$. That is define $S_N = \sum_{i=1}^{N} x_i$.

Definition I.4.3. *The series* $\sum_{i=1}^{\infty} x_i$ *is said to converge if the sequence of partial sums,* S_N, *converges (as a sequence) as* $N \to \infty$. *If* $\lim_{N\to\infty} S_N = a$, *then we write* $\sum_{i=1}^{\infty} x_i = a$.

Example I.4.5. Let $x_n = 1/3^n$ for $n = 1, 2, 3, \ldots$. The series formed by this sequence is called a geometric series and it is known that $\sum_{i=0}^{\infty} s^i = 1/(1-s)$ when $|s| < 1$. Hence for this example,

$$\sum_{n=1}^{\infty} \frac{1}{3^n} = \frac{1}{1 - \frac{1}{3}} - 1 = \frac{3}{2} - 1 = \frac{1}{2}.$$

The sequence of partial sums could have been used to find the value of the series. However in cases such as a geometric series, it is easier to evaluate the series directly.

One of the main mathematical problems that occurs in the remainder of this book is the problem of interchanging the operations of taking limits and infinite sums. Consider the following situation. Let a_{nk} be a doubly indexed sequence. That is the numbers in this sequence are given by

$$\begin{matrix} a_{11} & a_{12} & a_{13} & a_{14} & \cdots \\ a_{21} & a_{22} & a_{23} & a_{24} & \cdots \\ a_{31} & a_{32} & a_{33} & a_{34} & \cdots \\ \vdots & \vdots & \vdots & \vdots \end{matrix}$$

Assume that for each k the limit as $n \to \infty$ exists and equals a_k. That is, $\lim_{n \to \infty} a_{nk} = a_k$ for $k = 1, 2, \ldots$. For this situation is it necessarily true that

$$\lim_{n \to \infty} \sum_{k=1}^{\infty} a_{nk} = \sum_{k=1}^{\infty} a_k ? \tag{4.1}$$

In most calculus courses one learns that the limit of a sum is the sum of the limits if the limits exist. In this case the limits exist but the calculus result does *not* apply since there are infinitely many terms in the sum. To see how (4.1) could fail consider the following example.

Example I.4.6. Let $a_{nk} = 1/(n \cdot k)$ for $n = 1, 2, 3, \ldots$ and $k = 1, 2, 3, \ldots$. In this case $\lim_{n \to \infty} a_{nk} = 0$ for all k so $\sum_{k=1}^{\infty} a_k = 0$. However for every n we have

$$\sum_{k=1}^{\infty} a_{nk} = \sum_{k=1}^{\infty} \frac{1}{n \cdot k} = \frac{1}{n} \cdot \sum_{k=1}^{\infty} \frac{1}{k} = \infty.$$

Hence

$$\lim_{n \to \infty} \sum_{k=1}^{\infty} a_{nk} = \infty.$$

In view of this example some additional condition is needed before an interchange of lim and Σ can be done. Theorem I.4.1 provides such a sufficient condition. This theorem is called the Lebesgue Dominated Convergence Theorem. Most proofs of this theorem are given using the notation of the Lebesgue integral, and since an infinite sum is simply a special case of such an integral, the theorem is not usually proved directly for sums. However, since this result for sums will be used repeatedly in this book, a statement and proof for sums will be given.

Theorem I.4.1. (Lebesgue Dominated Convergence Theorem).

Let a_{nk} be a doubly indexed sequence of real numbers such that for each k, $\lim_{n \to \infty} a_{nk} = a_k$ exists. If there exists a sequence of non-negative numbers, b_k, such that for each k, $|a_{nk}| \leqslant b_k$ for $n = 1, 2, 3, \ldots$, and $\sum_{k=1}^{\infty} b_k < \infty$, then

$$\lim_{n \to \infty} \sum_{k=1}^{\infty} a_{nk} = \sum_{k=1}^{\infty} a_k.$$

Proof. In order to show that $\lim_{n \to \infty} \sum_{k=1}^{\infty} a_{nk} = \sum_{k=1}^{\infty} a_k$ the definition of a limit will be used. In particular, let $\epsilon > 0$ be given. We must show that there exists an integer, $N = N(\epsilon)$, such that $|\sum_{k=1}^{\infty} a_{nk} - \sum_{k=1}^{\infty} a_k| < \epsilon$ whenever $n \geqslant N$. Since for each k, $|a_{nk}| \leqslant b_k$ for all n, it follows that $|a_k| \leqslant b_k$ for each k. Now $\sum_{k=1}^{\infty} b_k < \infty$ so there exists an integer, $M = M(\epsilon)$ such that $\sum_{k=M+1}^{\infty} b_k < \epsilon/3$. This is possible since the tail of a convergent series goes to zero as $M \to \infty$. Fix M with the above property. Using the fact that $|x + y| \leqslant |x| + |y|$ we get

$$\left| \sum_{k=1}^{\infty} a_{nk} - \sum_{k=1}^{\infty} a_k \right| \leqslant \left| \sum_{k=1}^{M} a_{nk} - \sum_{k=1}^{M} a_k \right| + \sum_{k=M+1}^{\infty} |a_{nk}| + \sum_{k=M+1}^{\infty} |a_k|$$

$$\leqslant \sum_{k=1}^{M} |a_{nk} - a_k| + 2 \sum_{k=M+1}^{\infty} |b_k| < \sum_{k=1}^{M} |a_{nk} - a_k| + \frac{2\epsilon}{3}.$$

For each k we have $a_{nk} \to a_k$ so for $k = 1$ there exists $N_1 = N_1(\epsilon)$ such that $|a_{n1} - a_1| < \epsilon/3M$ for $n \geqslant N_1$. Similarly there exists $N_2 = N_2(\epsilon)$ such that $|a_{n2} - a_2| < \epsilon/3M$ for $n \geqslant N_2$. Continue this until an integer N_M is found such that $|a_{nM} - a_M| < \epsilon/3M$ for all $n \geqslant N_M$. Now let $N = \max\{N_1, N_2, \ldots, N_M\}$. Since there are only finitely many N_i's to consider, N is a finite number with the property that if $n \geqslant N$, then

$$\sum_{k=1}^{M} |a_{nk} - a_k| \leqslant \sum_{k=1}^{M} \frac{\epsilon}{3M} = \frac{\epsilon}{3}.$$

Hence whenever $n \geqslant N$ it follows that $|\sum_{k=1}^{\infty} a_{nk} - \sum_{k=1}^{\infty} a_k| < \epsilon$ so

$$\lim_{n \to \infty} \sum_{k=1}^{\infty} a_{nk} = \sum_{k=1}^{\infty} a_k. \quad \blacktriangle$$

In order to invoke this theorem as a justification for interchanging the operations of taking limits and infinite sums, the user must exhibit the dominating sequence $\{b_k\}$. The following example shows how this is done.

Example I.4.7. Let $a_{nk} = (-1)^k[(n+1)/n2^k]$. In this case $\lim_{n\to\infty} a_{nk} = (-1)^k/2^k$. Now we need to find a sequence, $\{b_k\}$, depending on k but not on n such that $|a_{nk}| \leqslant b_k$ and $\Sigma_{k=1}^{\infty} b_k < \infty$. Now $|(-1)[(n+1)/n]| \leqslant 2$ for $n = 1, 2, 3, \ldots$ so

$$|a_{nk}| \leqslant \frac{2}{2^k} = \frac{1}{2^{k-1}} \quad \text{and} \quad \sum_{k=1}^{\infty} \frac{1}{2^{k-1}} = \frac{1}{1-1/2} = 2.$$

Hence $b_k = 1/2^{k-1}$ for $k = 1, 2, 3, \ldots$ forms a dominating sequence so

$$\lim_{n\to\infty} \sum_{k=1}^{\infty} (-1)^k \frac{(n+1)}{n2^k} = \sum_{k=1}^{\infty} \frac{(-1)^k}{2^k}.$$

The reader is cautioned that, although this is a very powerful theorem, there are times when it cannot be used. If the dominating sequence cannot be exhibited, the theorem cannot be invoked.

Another operation that is closely related to the above is that of interchanging the order of summation for two infinite series. Again for finite sums it is well known that

$$\sum_{n=1}^{N} \sum_{k=1}^{K} a_{nk} = \sum_{k=1}^{K} \sum_{n=1}^{N} a_{nk}.$$

However, when the series become infinite this equality may fail as the following example shows.

Example I.4.8. Let $a_{nk} = (-1)^{nk}/n2^k$. First note that if n is even, then $\Sigma_{k=1}^{\infty}[(-1)^{nk}/2^k] = 1$ and if n is odd, then $\Sigma_{k=1}^{\infty}[(-1)^{nk}/2^k] = -\frac{1}{3}$. Hence

$$\sum_{n=1}^{\infty} \sum_{k=1}^{\infty} \frac{(-1)^{nk}}{n2^k} = \sum_{n=1}^{\infty} \frac{(-1)^n c_n}{n}$$

where $c_n = 1$ if n is even and $c_n = \frac{1}{3}$ if n is odd. The series $\Sigma_{n=1}^{\infty}[(-1)^n c_n/n]$ is conditionally convergent, so $\Sigma_{n=1}^{\infty}\Sigma_{k=1}^{\infty}[(-1)^{nk}/n2^k]$ is finite. However, if the two sums are interchanged, we get

$$\sum_{k=1}^{\infty} \sum_{n=1}^{\infty} \frac{(-1)^{nk}}{n2^k} = \sum_{k=1}^{\infty} \frac{(-1)^k d_k}{2^k}$$

where d_k is a finite negative number if k is odd and $d_k = +\infty$ if k is even. This series does not converge, so

$$\sum_{n=1}^{\infty} \sum_{k=1}^{\infty} a_{nk} \neq \sum_{k=1}^{\infty} \sum_{n=1}^{\infty} a_{nk}.$$

As in the case of interchanging limits and series, the theorem that yields sufficient conditions for interchanging two infinite series is usually stated and proved for integrals. Since our use of this theorem will involve sums only, it will be stated here with sums. However, no proof will be given.

Theorem I.4.2   (Fubini's Theorem).

Consider the double series $\sum_{n=1}^{\infty} \sum_{k=1}^{\infty} a_{nk}$. Any of the following conditions is sufficient to justify the interchange of the order of summation.

(i) $a_{nk} \geqslant 0$ for all k and n.

(ii) $\displaystyle\sum_{n=1}^{\infty} \sum_{k=1}^{\infty} |a_{nk}| < \infty.$

(iii) $\displaystyle\sum_{k=1}^{\infty} \sum_{n=1}^{\infty} |a_{nk}| < \infty.$

The next result we consider is again one that is usually stated and proved for integrals. In particular we are interested in the technique of integration by parts that is done in most calculus courses. However, as in the previous cases, we wish to apply this technique to sums rather than integrals. It might seem as if this small change should be easy since the sum is simply a special type of integral. In a sense this is true, but in actually writing down the result, special care must be taken.

In the case of integration by parts we recall that

$$\int_a^b f(x) g'(x) \, dx = f(b) g(b) - f(a) g(a) - \int_a^b g(x) f'(x) \, dx.$$

The analogous formula for sums will be referred to as summation by parts. Let $\{a_k\}_{k=1}^{\infty}$ and $\{b_k\}_{k=1}^{\infty}$ be two sequences of real numbers. Consider $\sum_{j=M}^{N} a_j [b_{j+1} - b_j]$. The a_j sequence plays the role of the function $f(x)$ above and the $(b_{j+1} - b_j)$ sequence plays the role of the function $g'(x)$ above. Hence one might guess that the summation by parts formula is

$$\sum_{j=M}^{N} a_j [b_{j+1} - b_j] = a_N b_N - a_M b_M - \sum_{j=M}^{N} b_j [a_{j+1} - a_j]. \tag{4.2}$$

This formula is almost correct except some corrections must be made regarding the subscripts and the range of summation. For example on the left-hand side of (4.2) the coefficient of a_M is $b_{M+1} - b_M$. On the right-hand side the coefficient of a_M is $b_M - b_M = 0$. In view of this we see that a naive substitution into the integration by parts formula does not yield a correct summation by parts formula. One reason for the difference is that when working with continuous functions it does not matter whether one considers $f(x)$ or $f(x-)$ since they are equal, but for sequences, $b_j \neq b_{j-1}$. By making appropriate adjustments to the summation by parts formula (4.2) we get the following correct formula:

$$\sum_{j=M}^{N} a_j [b_{j+1} - b_j] = a_N b_{N+1} - a_M b_M - \sum_{j=M+1}^{N} b_j [a_j - a_{j-1}]. \quad (4.3)$$

There are various forms in which this equation may appear. Hence the reader should not attempt to remember this particular form. Rather it is suggested that he remember that the equation comes from the integration by parts formula and that the subscripts and ranges of summation should be checked carefully.

The next definition we consider is not usually given in courses on advanced calculus but is frequently given in probability courses.

Definition I.4.4. *Let $\{a_k\}_{k=0}^{\infty}$ be a sequence of real numbers. The* generating function *determined by this sequence is defined by* $\mathcal{Q}(s) = \sum_{k=0}^{\infty} a_k s^k$.

If $\{a_k\}$ is a bounded sequence, then the series $\mathcal{Q}(s)$ certainly converges for $|s| < 1$. This is the type of sequence that will be encountered most frequently in this book since generally the $\{a_k\}$ sequence will be a sequence of probabilities and hence bounded by one.

Example I.4.9. Let $a_k = (\frac{1}{2})^k$ for $k = 0, 1, 2, \ldots$. The generating function determined by this sequence is

$$\mathcal{Q}(s) = \sum_{k=0}^{\infty} \left(\frac{s}{2}\right)^k = \frac{1}{1 - s/2} = \frac{2}{2 - s} \qquad \text{for } |s| < 1.$$

(In fact this example converges for $|s| < 2$ but the common radius of convergence of $|s| < 1$ for bounded sequences will be used throughout.)

The next topic in this section will be a statement and proof of Abel's theorem. This theorem serves two useful purposes at this point. First, the results of the theorem are needed later in the book and second, many of the ideas discussed in this section are used to prove this theorem.

Theorem I.4.3   (Abel's Theorem)

(i) If $\{a_k\}_{k=0}^{\infty}$ is a sequence of real numbers such that $\sum_{k=0}^{\infty} a_k$ converges, then

$$\lim_{s \to 1^-} \sum_{k=0}^{\infty} a_k s^k = \sum_{k=0}^{\infty} a_k.$$

(ii) If $\{a_k\}_{k=0}^{\infty}$ is a sequence of non-negative numbers such that

$$\lim_{s \to 1^-} \sum_{k=0}^{\infty} a_k s^k = a \leqslant \infty,$$

then $\sum_{k=0}^{\infty} a_k = a$. (Note that $\lim_{s \to 1^-}$ means the limit is taken as s approaches 1 from the left.)

Proof. For part (i) it must be shown that given $\epsilon > 0$ there exists $\delta > 0$ such that $|\sum_{k=0}^{\infty} a_k s^k - \sum_{k=0}^{\infty} a_k| < \epsilon$ whenever $0 < 1 - s < \delta$. Note that Definition I.4.1' is used here. Now

$$\left| \sum_{k=0}^{\infty} a_k (s^k - 1) \right| \leqslant \left| \sum_{k=0}^{M} a_k (s^k - 1) \right| + \left| \sum_{k=M+1}^{\infty} a_k (s^k - 1) \right| \qquad (4.4)$$

so the two terms on the right-hand side will be considered separately. The first step is to choose M so large that the second term on the right-hand side of (4.4) is less than $\epsilon/2$ independently of s. [If the hypothesis of this theorem were $\sum_{k=0}^{\infty} |a_k| < \infty$, then the choice of M would be trivial since $|s^k - 1| < 1$. That is,

$$\left| \sum_{k=M+1}^{\infty} a_k (s^k - 1) \right| \leqslant \sum_{k=M+1}^{\infty} |a_k| |s^k - 1| \leqslant \sum_{k=M+1}^{\infty} |a_k| < \epsilon/2$$

for M sufficiently large. In fact, Lebesgue's dominated convergence theorem could be used to show part (i) directly under the assumption that $\sum_{k=0}^{\infty} |a_k| < \infty$. Unfortunately the assumption is that $\sum_{k=0}^{\infty} a_k$ converges, not that it converges absolutely.] First choose $M = M(\epsilon)$ so large that $|\sum_{k=m}^{\infty} a_k| < \epsilon/4$ for all $m \geqslant M$. Then

$$\sum_{k=M+1}^{\infty} a_k (s^k - 1) = \sum_{k=M+1}^{\infty} (1 - s^k) \left[\sum_{j=k+1}^{\infty} a_j - \sum_{j=k}^{\infty} a_j \right]. \qquad (4.5)$$

Using summation by parts we get that (4.5) is equal to

$$\left(\sum_{j=k}^{\infty} a_j\right)(1-s^k)\bigg|_{k=M+1}^{\infty} - \sum_{k=M+2}^{\infty} \left(\sum_{j=k}^{\infty} a_j\right)(s^{k-1}-s^k)$$

$$\leqslant 0 + \left|\sum_{j=M+1}^{\infty} a_j\right| |1-s^{M+1}| + \sum_{k=M+2}^{\infty} \left|\sum_{j=k}^{\infty} a_j\right| (s^{k-1}-s^k)$$

$$\leqslant \epsilon/4 + (\epsilon/4)s^{M+1} \leqslant \epsilon/2 \qquad \text{for} \quad 0 < s < 1.$$

Now with M fixed so the second term of (4.4) is less than $\epsilon/2$, the first term can be made less that $\epsilon/2$ by taking s sufficiently close to 1. That is there exists $\delta > 0$ such that each of the $M+1$ terms in $\sum_{k=0}^{M} a_k(s^k - 1)$ can be made less than $\epsilon/2(M+1)$ when $s > 1 - \delta$. Hence $|\sum_{k=0}^{\infty} a_k(s^k - 1)| < \epsilon$ for $0 < 1 - s < \delta$ and part (i) is proved.

For part (ii) the two cases $a < \infty$ and $a = \infty$ are considered separately. If $a = \infty$, the proof is quite easy since $\sum_{k=0}^{\infty} a_k s^k \leqslant \sum_{k=0}^{\infty} a_k$ for $s \in (0, 1)$ and $a_k \geqslant 0$. Hence $\sum_{k=0}^{\infty} a_k = \infty$ if $\lim_{s \to 1^-} \sum_{k=0}^{\infty} a_k s^k = \infty$. If $a < \infty$, then

$$\sum_{k=0}^{n} a_k = \lim_{s \to 1^-} \sum_{k=0}^{n} a_k s^k \leqslant a$$

for all n. Hence $\sum_{k=0}^{n} a_k$ forms a bounded monotone sequence as $n \to \infty$, so $\sum_{k=0}^{\infty} a_k$ converges to some real number that we will call a' (see Exercise 6). By part (i) we know $a' = \lim_{s \to 1^-} \sum_{k=0}^{\infty} a_k s^k$ so it follows that $a' = a$. ▲

The proof of the next theorem is another good exercise in the application of the concept of a limit. The proof of this theorem uses a technique that has already been used in the proof of Lebesgue's dominated convergence theorem and Abel's theorem, namely the decomposition of a sum into two parts in such a way that both parts can be shown to be small (see Equation 4.7). One reason for including the proof of this theorem is that this technique may be useful to the reader in solving other limit problems involving series. A more cogent reason for including the next theorem is that it is needed repeatedly later in the book in order to prove important results.

Theorem I.4.4.

If $\{a_k\}_{k=0}^{\infty}$ is a sequence of numbers such that $\sum_{k=0}^{\infty} |a_k| < \infty$ and $\sum_{k=0}^{\infty} a_k = a$, and if $\{b_k\}_{k=0}^{\infty}$ is a sequence of numbers such that $\lim_{n \to \infty} b_n = b$, then

$$\lim_{n \to \infty} \sum_{k=0}^{n} a_k b_{n-k} = ab.$$

Proof.   Let $\epsilon > 0$ be given. Consider

$$\left| \sum_{k=0}^{n} a_k b_{n-k} - ab \right| = \left| \sum_{k=0}^{n} a_k b_{n-k} - \sum_{k=0}^{\infty} a_k b \right|$$

$$\leqslant \left| \sum_{k=0}^{n} a_k(b_{n-k} - b) \right| + \left| \sum_{k=n+1}^{\infty} a_k b \right|. \quad (4.6)$$

The second term of (4.6) will certainly be small for n sufficiently large, since $\sum_{k=0}^{\infty} |a_k| < \infty$ implies $|\sum_{k=n+1}^{\infty} a_k|$ must be small for large n. For this reason we may focus our attention on the first term of (4.6), namely $\sum_{k=0}^{n} a_k(b_{n-k} - b)$.

We have, by hypothesis, that $b_n \to b$, so when $(n-k)$ is sufficiently large, $|b_{n-k} - b|$ will be small. However since the range of summation is $k = 0, 1, \ldots, n$, no matter how large n is chosen, we will have $n - k$ "small" for part of the range of summation. Fortunately when $n - k$ is small, k will be large and consequently a_k will be small. Hence one of the two factors, a_k or $(b_{n-k} - b)$ will be small for all k. We can exploit this fact as follows. Express the first sum in (4.6) as

$$\sum_{k=0}^{n} a_k(b_{n-k} - b) = \sum_{k=0}^{M} a_k(b_{n-k} - b) + \sum_{k=M+1}^{n} a_k(b_{n-k} - b). \quad (4.7)$$

Since $b_n \to b$, the sequence of numbers $\{b_n\}$ must be bounded in absolute value (Exercise 7). Let B denote this bound. Next note that since $\sum_{k=1}^{\infty} |a_k| < \infty$, there exists $M = M(\epsilon)$ such that $\sum_{k=M+1}^{\infty} |a_k| < \epsilon/(6 \cdot B)$. For this choice of M, it follows that for all $n > M$ that

$$\left| \sum_{k=M+1}^{n} a_k(b_{n-k} - b) \right| \leqslant \sum_{k=M+1}^{n} |a_k| |b_{n-k} - b| \leqslant 2B \sum_{k=M+1}^{n} |a_k|$$

$$\leqslant 2B \sum_{k=M+1}^{\infty} |a_k| < \frac{\epsilon}{3}.$$

Considering the first term on the right-hand side of (4.7), we see that the smallest value of the subscript $n - k$ will be $n - M$. In view of the remarks preceding (4.7), we know that there exists $N = N(\epsilon, M)$ such that $|b_{n-M} - b| < \epsilon/(3(\sum_{k=0}^{M} |a_k|))$ for all $n \geqslant N$. Hence for all $n \geqslant N$,

$$\left| \sum_{k=0}^{M} a_k(b_{n-k} - b) \right| \leqslant \sum_{k=0}^{M} |a_k| |b_{n-k} - b| < \frac{\epsilon}{3}.$$

Consequently the left-hand side of (4.7) is bounded in absolute value by $2\epsilon/3$. Furthermore, by construction we have that $N = N(\epsilon, M) > M$, hence for $n \geqslant N$ it follows that

$$\left| \sum_{k=n+1}^{\infty} a_k b \right| \leqslant B \sum_{k=n+1}^{\infty} |a_k| < B\left(\frac{\epsilon}{(6 \cdot B)} \right) = \frac{\epsilon}{6}.$$

Therefore we have from (4.6) that for all $n \geqslant N$,

$$\left| \sum_{k=0}^{n} a_k b_{n-k} - ab \right| \leqslant \frac{2\epsilon}{3} + \frac{\epsilon}{6} < \epsilon.$$

This concludes the proof of the theorem. ▲

Theorem I.4.4 is also true when $a = \infty$ and $b \neq 0$. The proof of this case is left as an exercise (Exercise 11).

We would again emphasize that in the proof of this theorem we utilized the technique of partitioning a sum into two parts in such a way that one of the hypotheses, $\sum_{k=0}^{\infty} |a_k| < \infty$ or $\lim_{n \to \infty} b_n = b$, could be used to show each part to be small.

In Chapter V we will find it useful to use the concept of infinite products. In particular we will consider $\lim_{n \to \infty} \prod_{k=0}^{n} (1 - a_k)$, where $0 \leqslant a_k \leqslant 1$.

Definition I.4.5. *Let $\{a_k\}_{k=0}^{\infty}$ be a sequence of numbers satisfying $0 \leqslant a_k \leqslant 1$. The infinite product $\prod_{k=0}^{\infty}(1 - a_k)$ is said to converge if*

$$\lim_{n \to \infty} \prod_{k=0}^{n} (1 - a_k) = \alpha > 0.$$

If $\lim_{n \to \infty} \prod_{k=0}^{n}(1 - a_k) = 0$, then the product is said to diverge to zero.

At first glance the reader might be a bit dismayed at having us refer to an infinite product as "divergent" when the limit is zero. However, since for all k it is true that $0 \leqslant a_k \leqslant 1$ (and consequently $0 \leqslant 1 - a_k \leqslant 1$) we can give some justification for this terminology by considering logs. Since we have

$$\log \prod_{k=0}^{n} (1 - a_k) = \sum_{k=0}^{n} \log(1 - a_k),$$

the "divergence" of $\prod_{k=0}^{n}(1 - a_k)$ to zero as $n \to \infty$ is equivalent to the divergence of $\sum_{k=0}^{n} \log(1 - a_k)$ to minus infinity.

There is one particular aspect of this definition which is rather unsatisfying and that is the effect that a finite number of elements (indeed a single element) can have on the convergence or divergence of the infinite product. Specifically, if a single factor is zero (equivalently if $a_k = 1$ for some k) then the infinite product is zero. (Why?) When considering convergence of an infinite series, we find that the behavior of the *tail* of the series is of crucial importance and we would like to have a similar property hold for infinite products. This can be accomplished by revising Definition I.4.5 as follows.

Definition I.4.6. *Let* $\{a_k\}_{k=0}^{\infty}$ *be a sequence of numbers satisfying* $0 \leqslant a_k \leqslant 1$. *The infinite product is said to converge if for some m,*

$$\lim_{n \to \infty} \prod_{k=m}^{n} (1 - a_k) = \alpha > 0.$$

If for all m, $\lim_{n \to \infty} \prod_{k=m}^{n} (1 - a_k) = 0$, *the infinite product is said to diverge to zero.*

The following theorem gives necessary and sufficient conditions for the divergence of such infinite products.

Theorem I.4.5.

Let $\{a_k\}_{k=0}^{\infty}$ be a sequence of numbers satisfying $0 \leqslant a_k \leqslant 1$. The infinite product $\prod_{k=0}^{\infty}(1 - a_k)$ diverges to zero if and only if $\sum_{k=0}^{\infty} a_k = \infty$.

Proof. Assume that $\sum_{k=0}^{\infty} a_k = \infty$. If infinitely many of the a_k's are equal to one, then the infinite product certainly diverges to zero. Therefore we will consider the case where $0 \leqslant a_k < 1$ for $k \geqslant m$ and for some $m \geqslant 0$. It can be shown (Exercise 8) that for $x \in [0, 1)$, $1 - x \leqslant e^{-x}$. This implies that

$$0 < \prod_{k=m}^{n} (1 - a_k) \leqslant \exp\left(-\sum_{k=m}^{n} a_k \right). \tag{4.8}$$

Taking the limit as $n \to \infty$ of both sides of (4.8) and using the fact that $\sum_{k=m}^{\infty} a_k = \infty$, we get

$$0 \leqslant \lim_{n \to \infty} \prod_{k=m}^{n} (1 - a_k) \leqslant \lim_{n \to \infty} \exp\left(-\sum_{k=m}^{n} a_k \right) = 0.$$

The converse can be proven by contradiction. Assume that the infinite product diverges to zero. We wish to show that $\sum_{k=0}^{\infty} a_k = \infty$. Assume that this is not true, that is, assume $\sum_{k=0}^{\infty} a_k = a < \infty$. Then given $\epsilon > 0$ there exists $M = M(\epsilon)$ such that $0 \leqslant \sum_{k=M}^{\infty} a_k < \epsilon$. In Exercise 9 the reader will be asked to show that

$$\prod_{k=M}^{n} (1 - a_k) = (1 - a_M)(1 - a_{M+1}) \cdots (1 - a_n) \geqslant 1 - \sum_{k=M}^{n} a_k.$$

Using this result we have, by the way that M was chosen, that

$$\prod_{k=M}^{n} (1 - a_k) \geqslant 1 - \sum_{k=M}^{n} a_k \geqslant 1 - \sum_{k=M}^{\infty} a_k > 1 - \epsilon \qquad (4.9)$$

for all n. It is easy to see that $\prod_{k=M}^{n}(1 - a_k)$ is monotone decreasing as n increases so (by Exercise 6) $\lim_{n \to \infty} \prod_{k=M}^{n}(1 - a_k)$ exists. Further, by (4.9) this limit is bounded away from zero, hence $\prod_{k=0}^{\infty}(1 - a_k)$ converges. Since this contradicts our original assumption, it must be that $\sum_{k=0}^{\infty} a_k = \infty$. ▲

The last topic that we consider in this section of mathematical review is that of a norm. In Chapter V we will define a norm on a set of vectors and on a set of matrices and will make use of the properties of norms in proving certain important theorems. Here we give a brief discussion of those properties that will be needed later.

Generally a norm is defined on an arbitrary linear space, but for our purposes it is not necessary to give the definition in its fullest generality. Rather we will define the two particular linear spaces with which we will be concerned and consider norms on these spaces only.

Definition I.4.7. *Define $V_n = \{\mathbf{x} = (x_1, x_2, \ldots, x_n) : x_i \text{ is a real number}\}$. V_n is called the set of real n-dimensional vectors.*

The following properties hold for elements of V_n:

(i) If $\mathbf{x}$ and $\mathbf{y}$ belong to V_n and if α and β are real numbers then it is true that $\alpha \mathbf{x} + \beta \mathbf{y}$ belongs to V_n. That this is true is easy to see since

$$\alpha \mathbf{x} + \beta \mathbf{y} = \alpha (x_1, x_2, \ldots, x_n) + \beta (y_1, y_2, \ldots, y_n)$$

$$= (\alpha x_1, \alpha x_2, \ldots, \alpha x_n) + (\beta y_1, \beta y_2, \ldots, \beta y_n)$$

$$= (\alpha x_1 + \beta y_1, \alpha x_2 + \beta y_2, \ldots, \alpha x_n + \beta y_n)$$

which is in V_n. In words this says that V_n is closed under addition and scalar multiplication.

(ii) The vector $\varphi = (0, 0, \ldots, 0)$, called the null or zero vector, has the property that for all $\mathbf{x}$ in V_n, $\mathbf{x} + \varphi = \mathbf{x}$ and φ is the only vector with this property.

Conditions (i) and (ii) must hold in order for V_n to be called a linear space. In fact there are other conditions that must be satisfied as well in order for V_n to be called a linear space. However, since these other properties are not of immediate concern to us, we do not discuss them further. The interested reader may consult any text on linear algebra for a more complete discussion of linear spaces.

Definition I.4.8. *Define* $M_n = \{n \times n$ *matrices with real entries*$\}$.

It is easy to show that M_n is closed under addition and scalar multiplication and has a unique zero element (Exercise 10). Although we have not attempted to prove it, it is true that both V_n and M_n are linear spaces. We will now define a norm on a linear space.

Definition I.4.9. *A norm on a linear space, L, is a function that maps each X in L into a non-negative real number, denoted by $\|X\|$, in such a way that*

1. For all X and Y in L, $\|X + Y\| \leqslant \|X\| + \|Y\|$. (This inequality is referred to as the triangle inequality for norms.)
2. For all X in L and real numbers α, $\|\alpha X\| = |\alpha| \cdot \|X\|$.
3. $\|X_0\| = 0$ if and only if X_0 is the zero element of L.

Example I.4.10. If we consider the linear space V_3, we can show that the following function is in fact a norm:

$$\|\mathbf{x}\| = \|(x_1, x_2, x_3)\| = |x_1| + |x_2| + |x_3|.$$

Since the absolute value of a real number is a non-negative real number, it suffices to show that conditions 1 to 3 hold:

1. For any real numbers, a and b, we have $|a + b| \leqslant |a| + |b|$, hence

$$\|\mathbf{x} + \mathbf{y}\| = \|(x_1 + y_1, x_2 + y_2, x_3 + y_3)\|$$

$$= |x_1 + y_1| + |x_2 + y_2| + |x_3 + y_3|$$

$$\leqslant |x_1| + |y_1| + |x_2| + |y_2| + |x_3| + |y_3| = \|\mathbf{x}\| + \|\mathbf{y}\|.$$

2. $\|\alpha \mathbf{x}\| = \|(\alpha x_1, \alpha x_2, \alpha x_3)\|$

$$= |\alpha x_1| + |\alpha x_2| + |\alpha x_3| = |\alpha| |x_1| + |\alpha| |x_2| + |\alpha| |x_3|$$

$$= |\alpha|(|x_1| + |x_2| + |x_3|) = |\alpha| \|\mathbf{x}\|$$

3. The zero element is $\varphi = (0,0,0)$, so $\|\varphi\| = |0| + |0| + |0| = 0$

Conversely if $\|x\| = |x_1| + |x_2| + |x_3| = 0$ then it must be true that $x_1 = x_2 = x_3 = 0$.

Other examples of norms on the linear spaces V_n and M_n are given in the exercises. These examples illustrate the important fact that many *different* norms can be defined on a single linear space.

One important reason for defining a norm on a linear space L is that it can be used to define the "distance" between two elements of L. More explicitly, if X and Y belong to L then the distance between X and Y can be defined to be $d(X,Y) = \|X - Y\|$. Further, the introduction of the concept of distance allows for consideration of convergence of a sequence of elements in L.

Definition I.4.10. *Let L be a linear space and let X and X_n, $n = 1, 2, \ldots$ be elements of L. We say that X_n converges in norm to X if $\lim_{n \to \infty} \|X_n - X\| = 0$.*

Observe that the norm plays the same role for elements of L as the absolute value plays for real numbers in the definition of convergence of real numbers. It is very important to remember, however, that the convergence in norm of X_n to X does in fact depend on which of the different possible norms is used. The following example illustrates this dependence.

Example I.4.11. Let L be the linear space of continuous real valued functions on $[0, \infty)$. Define two norms on this space as follows. For f belonging to L, define

(i) $\|f\|_1 = \displaystyle\int_0^\infty |f(x)| \, dx$

(ii) $\|f\|_2 = \displaystyle\sup_{x \in [0, \infty)} |f(x)|.$

The sequence of functions,

$$f_n(x) = \begin{cases} \dfrac{1}{n} & \text{for } x \text{ in } [0, n] \\ 0 & \text{otherwise} \end{cases}$$

converges to zero using the norm $\|\cdot\|_2$, but not using the norm $\|\cdot\|_1$.

The problem of a sequence converging relative to one norm but not another will be discussed further in Chapter V when we consider pointwise and norm convergence of infinite stochastic matrices.

EXERCISES

1. Let $a_n = (n+1)/2^n$ for $n = 0, 1, 2, \ldots$. Find the generating function, $\mathcal{Q}(s)$, determined by this sequence.

2. Let X be a random variable with $P[X = k] = (\frac{1}{2})^k$ for $k = 1, 2, 3, \ldots$. Find the generating function determined by $\{(\frac{1}{2})^k\}_{k=1}^\infty$. This is often called the probability generating function of X. Use this to find the mean and variance of X.

3. Find $\sum_{k=0}^\infty s^k$, $\sum_{k=0}^\infty k \cdot s^k$, and $\sum_{k=0}^\infty k^2 s^k$ for $|s| < 1$.

4. Determine which of the following functions on V_n is a norm:

   (a) $\|x\| = \sup_{1 \leqslant i \leqslant n} |x_i|$ (c) $\|x\| = (x_1^2 + x_2^2 + \cdots + x_n^2)^{1/2}$

   (b) $\|x\| = |x_1|$ (d) $\|x\| = \left| \sum_{i=1}^n x_i \right|$.

5. Determine which of the following functions on M_n is a norm:

   (a) $\|M\| = \sum_{j=1}^n \sum_{i=1}^n |m_{ij}|$ (c) $\|M\| = \sum_{i=1}^n m_{ii}$

   (b) $\|M\| = \left| \sum_{j=1}^n \sum_{i=1}^n m_{ij} \right|$ (d) $\|M\| = \sup_{\substack{1 \leqslant i \leqslant n \\ 1 \leqslant j \leqslant n}} |m_{ij}|$

6. Prove that a bounded, monotone sequence of real numbers converges. That is, if $b_k \leqslant b_{k+1}$ for $k = 1, 2, 3, \ldots$ and $|b_k| \leqslant M$ for all k, then $\lim_{k \to \infty} b_k$ exists.

7. Prove that a convergent sequence of real numbers, $\{b_n\}$, is bounded in absolute value.

8. Prove that $1 - x \leqslant e^{-x}$ for $0 \leqslant x < 1$.

9. Prove that if $0 \leqslant a_k < 1$ for $k = 1, 2, \ldots, n$ then $(1 - a_1)(1 - a_2) \ldots (1 - a_n) \geqslant 1 - \sum_{k=1}^n a_k$. (Hint: Use induction on the number of factors.)

10. Show that M_n is closed under addition and scalar multiplication. Also show that M_n has a unique zero element.

11. Prove that Theorem I.4.4 holds when $a = \infty$ as long as $b \neq 0$.

*12. Let $\{a_n\}_{n=1}^{\infty}$ be a sequence of numbers that converge to a_0 as $n\to\infty$. Show that

$$\lim_{N\to\infty} \frac{1}{N} \sum_{n=1}^{N} a_n = a_0.$$

[Terms of the form $(1/N)\sum_{n=1}^{N} a_n$ are called Cesaro averages of $\{a_n\}$.]

*13. Let $\{a_n\}_{n=0}^{\infty}$ be a sequence of non-negative numbers such that $a_n/\sum_{k=0}^{n} a_k \to 0$ as $n\to\infty$. Let $\{b_n\}_{n=0}^{\infty}$ be a sequence of numbers converging to b. Show that

$$\frac{b_0 a_n + b_1 a_{n-1} + \cdots + b_n a_0}{\displaystyle\sum_{k=0}^{n} a_k} \to b \qquad \text{as } n\to\infty.$$

CHAPTER II

Fundamental Concepts of Markov Chains

DEFINITIONS AND PROPERTIES

In Chapter I it was shown how the starting vector and transition matrix for a discrete-time stationary Markov chain could be used to find the distribution of the chain after n steps. For chains with many states the problem of calculating $\mathbf{a}_n = \mathbf{a}_0 P^n$ becomes quite tedious, especially for large n. In some cases the vector $\mathbf{a}_n$ converges to a fixed vector, $\boldsymbol{\pi}$, which is independent of $\mathbf{a}_0$. The limit vector, $\boldsymbol{\pi}$, is called the "long run distribution" or the "invariant distribution." The purpose of this chapter is to develop the terminology that will be needed in Chapter III to study the long run distribution.

As indicated above, not every discrete-time stationary Markov chain has the property that $\mathbf{a}_n$ converges independently of $\mathbf{a}_0$. Hence a further reduction of this class of processes is required. The following definitions will facilitate this reduction.

Definition II.1.1. *A subset, C, of the state space, S, is called* closed *if $p_{ik} = 0$ for all $i \in C$ and $k \notin C$. If a closed set consists of a single state, then that state is called an absorbing state.*

The above definition should not be confused with the concept of a closed set in analysis. Recall that a subset of the real line is closed if it contains all of its limit points. In the case of Markov chains, closure does not refer to inclusion of a boundary, but rather to the impossibility of escaping. That is, a subset C of S is closed if once the chain enters C, it can never leave C.

Example II.1.1. Consider a chain with five states ($S = \{1, 2, 3, 4, 5\}$). Let

43

the transition matrix be given by

$$P = \begin{bmatrix} \frac{1}{2} & 0 & 0 & \frac{1}{2} & 0 \\ \frac{1}{2} & 0 & \frac{1}{3} & 0 & \frac{1}{6} \\ 0 & 0 & 1 & 0 & 0 \\ 1 & 0 & 0 & 0 & 0 \\ 0 & 1 & 0 & 0 & 0 \end{bmatrix}.$$

As usual the ith row and column correspond to the ith state. Since $p_{3,3}=1$ we see that once the particle enters state 3, it will remain there. The single state 3 is an absorbing state, but it is not the only closed set for this example. Note that S itself is trivially a closed set. To find other closed sets, it sometimes is helpful to make a diagram as in Figure II.1.1. An arrow is drawn from state i to state j if $p_{ij} > 0$, that is, if it is possible to go directly from state i to state j. The value of p_{ij} is given next to the corresponding arrow. [Such diagrams are called directed graphs or digraphs. See Anderson (1970).]

From the figure it is easy to see that there are no arrows leaving the two sets $\{1,4\}$ and $\{3\}$. These two subsets of S are closed. Actually according to the definition the subset $\{1,3,4\}$ is also a closed set. However, further definitions will lead us to consider $\{1,4\}$ and $\{3\}$ as separate.

A simple illustration of an absorbing state is given by a modification of the mouse in the maze example. Say a "trap" for the mouse is placed in a given room—say room 9—and assume the trap is sure to go off when the mouse enters the room. In this case, it would be impossible for the mouse to leave the ninth state, which is an absorbing state. The imaginative reader can visualize a maze constructed in such a way that a certain set of states would form a closed set.

The use of a digraph such as that given in Figure II.1.1 is a standard

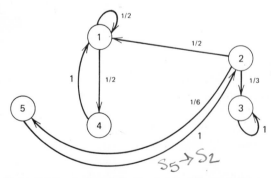

Figure II.1.1. A digraph for the matrix of Example II.1.1

technique in graph theory as well as Markov chains (see Anderson 1970). The reader may find this approach helpful in solving various Markov chain problems throughout this book.

Proposition II.1.1. If C is a closed set with $i \in C$ and $k \notin C$, then $p_{ik}^{(n)} = 0$ for all n.

Proof. The proof is left to the reader (Exercise 7).

Definition II.1.2. *A Markov chain is called* irreducible *if there exists no nonempty closed set other than S itself. If S has a proper closed subset, it is called reducible.*

Definition II.1.3. *Two states, i and j, are said to* intercommunicate *if for some $n \geq 0$, $p_{ij}^{(n)} > 0$ and for some $m \geq 0$, $p_{ji}^{(m)} > 0$.*

This definition says that it is possible for the chain to go from i to j in n steps and it is possible to go from j to i in m steps. The integers m and n need not be the same.

Example II.1.2. Consider a Markov chain with the following transition matrix:

$$P = \begin{bmatrix} 0 & 1 & 0 \\ 0 & 0 & 1 \\ 1 & 0 & 0 \end{bmatrix}.$$

If $S = \{1, 2, 3\}$ then states 1 and 2 intercommunicate since $p_{12}^{(1)} > 0$ and $p_{21}^{(2)} > 0$. In fact all the states in S intercommunicate and so by Theorem II.1.1 below, it follows that the chain is irreducible.

Theorem II.1.1.

A Markov chain is irreducible if and only if all pairs of states intercommunicate.

Proof. Assume the chain is irreducible. Define $C_j = \{i : p_{ij}^{(n)} = 0$ for all $n \geq 0\}$. That is, C_j is the set of all states from which state j cannot be reached. The set C_j is a closed subset of S. To prove C_j is closed it must be shown that if $i \in C_j$ and $k \notin C_j$, then $p_{ik} = 0$. Now if $k \notin C_j$ then for some $m \geq 0$ it follows that $p_{kj}^{(m)} > 0$. If p_{ik} were positive, then $p_{ij}^{(m+1)} = \sum_{l \in S} p_{il} p_{lj}^{(m)} \geq p_{ik} p_{kj}^{(m)} > 0$, which implies $i \notin C_j$. This contradiction leads us to conclude that $p_{ik} = 0$ for all $i \in C_j$, $k \notin C_j$, so C_j is closed. The only nonempty closed subset of an irreducible chain is S so $C_j = S$ or $C_j = \varnothing$. However $j \not\in C_j$ since $p_{jj}^{(0)} = 1$. Therefore $C_j = \varnothing$ which means j can be reached from all states. Since j is an arbitrary state we have that all states intercommunicate.

Conversely assume that all states intercommunicate. Let C be a nonempty closed set in S. If $j \in C$, then for an arbitrary state $i \in S$, there exists an n_i such that $p_{ji}^{(n_i)} > 0$. Since state i can be reached from state $j \in C$, it follows that $i \in C$. But i was an arbitrary state in S so $C = S$. Therefore the chain is irreducible. ▲

Returning to Example II.1.2 we see that the chain corresponding to this transition matrix is not only irreducible, but it has some additional noticeable characteristics. The chain moves among the states $\{1, 2, 3\}$ in a deterministic and periodic manner. While the class of Markov chains that move in a deterministic manner is not of general interest, the question of whether or not there is some period to the motion of a chain is of great importance. Before giving a formal definition of a periodic chain, an intuitive discussion of the concept will be given.

Most people have some feeling for what periodic means. Specifically, most people think of periodic behavior as being exhibited when something occurs at regular, predictable time or space intervals. A familiar example would be the chiming of a clock, which, barring mechanical breakdown, chimes once every 30 minutes, say, and most people would agree that the period is 30 minutes. Now consider a clock that *may* chime on the half-hour. In particular, say we have constructed a clock so that on every half-hour the clock will chime with probability p and will fail to chime with probability $1-p$. This situation is clearly different from having a clock chime every 30 minutes, yet since the clock *may* chime at 30-minute intervals, we could still say that the period is 30 minutes. Going somewhat further, if the clock were set so that the first time it could chime was at time zero, and if we considered the set $T = \{t : \text{clock may chime at time } t\} = \{0, 30, 60, 90, \ldots\}$, we would see that the period is the greatest common divisor of the set T. (Recall that if S is a set of non-negative integers, the greatest common divisor of the set S is the positive number, λ, with the following properties: (i) λ divides all the numbers in S, (ii) If λ' divides all the numbers in S, then λ' divides λ. (See Exercise 1.))

With this discussion in mind, consider the following Markov chain, which is described by the digraph in Figure II.1.2. In order to find the period for state 1, consider the times that, starting from state 1, the chain may return to state 1. The possible times are $4, 6, 8, 10, 12, \ldots$. If the time 2 were included in the sequence, we would certainly define the period to be 2. Actually, it is not unreasonable to define the period to be 2 in spite of the fact that the time 2 is not a possible time of return. Specifically, we define periodicity as follows:

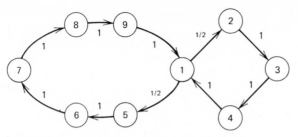

Figure II.1.2.   Digraph for a periodic Markov chain.

Definition II.1.4.   *State j has* period *d if the following two conditions hold:*

(i) $p_{jj}^{(n)} = 0$ unless $n = md$ for some positive integer m and
(ii) d is the largest integer with property (i).

State j is called aperiodic *when $d = 1$.*

Let us reconsider the example given by Figure II.1.2 in view of this definition. It is clear that $p_{11}^{(n)} = 0$ for n odd. Hence by part (i) of the definition, the period is even. However at this point the period could be 2 or 4 or 6 or larger. Now 4, for example, does not satisfy condition (i) since $p_{11}^{(n)}$ can be positive for n not equal to a multiple of 4 (for example, $p_{11}^{(6)} = \frac{1}{2}$). In fact the only even integer that satisfies condition (i) is 2, so 2 is the largest and hence the period is 2.

Example II.1.3.   Let $\{X_n\}$ be a discrete-time stationary Markov chain with state space $\{1,2,3,4\}$ and transition matrix

$$P = \begin{bmatrix} 0 & 1 & 0 & 0 \\ 0 & 0 & 1 & 0 \\ 0 & 0 & 0 & 1 \\ 1 & 0 & 0 & 0 \end{bmatrix}.$$

For this chain $p_{11}^{(n)} = 0$ unless n is even. Hence by part (i) of the definition the period is 2 or 4 or 6 or larger. Condition (i) is satisfied by the even integers 2 and 4. Hence by part (ii) of the definition, the period of this chain is 4. This example shows the need for the second part of the definition.

The reader may have noticed that the period of state i is somehow related to $\{n : p_{ii}^{(n)} > 0\}$. The following theorem establishes the relationship and gives a second characterization of the period of a state.

Theorem II.1.2.

State j has period d if and only if d is the greatest common divisor of all those n's for which $p_{jj}^{(n)} > 0$ (that is, $d = \text{G.C.D.} \{ n : p_{jj}^{(n)} > 0 \}$).

Proof. Assume j has period d. This implies that $p_{jj}^{(n)} = 0$ unless n is a multiple of d. (In this book the term multiple refers to integral multiple.) Hence those n for which $p_{jj}^{(n)} > 0$ must be multiples of d so G.C.D. $\{ n : p_{jj}^{(n)} > 0 \}$ is a multiple of d. Now if G.C.D. $\{ n : p_{jj}^{(n)} > 1 \} = kd$ then $p_{jj}^{(n)} = 0$ unless $n = m(kd)$. But by part (ii) of the definition of periodicity, d is the largest integer with this property so $k = 1$.

On the other hand assume $d = \text{G.C.D.} \{ n : p_{jj}^{(n)} > 0 \}$. This implies that d divides $\{ n : p_{jj}^{(n)} > 0 \}$ so $p_{jj}^{(n)} = 0$ unless $n = md$. Now in order to show d is the largest integer with this property, assume $d' > d$ and $p_{jj}^{(n)} = 0$ unless $n = md'$. Then d' divides $\{ n : p_{jj}^{(n)} > 0 \}$. This contradicts the assumption that d was the *greatest* common divisor. Hence d is the period of state j. ▲

The period of state j is concerned with the times at which the chain might return to state j. The following example shows that while it is *possible* for the chain to return to a state at certain times, it may in fact never return!

Example II.1.4. Let X_n be a Markov chain with state space $\{ 1, 2, 3, 4 \}$ and transition matrix

$$
P = \begin{bmatrix}
0 & 1 & 0 & 0 \\
\frac{1}{2} & 0 & \frac{1}{2} & 0 \\
0 & 0 & 0 & 1 \\
0 & 0 & 1 & 0
\end{bmatrix}
$$

The reader can verify that each state has period 2 and that since states 3 and 4 form a closed set, once the particle goes from state 2 to state 3 it will never again return to either state 1 or state 2.

In order to make the above notion explicit, we give the following definitions.

Definition II.1.5. *Let $f_{ij}^{(n)}$ denote the probability that the first visit to state j from state i occurs at time n. That is,*

$$
f_{ij}^{(n)} = P[X_{n+k} = j | X_{n+j-1} \neq j, X_{n+k-2} \neq j, \ldots, X_{k+1} \neq j, X_k = i].
$$

If $i = j$ we refer to $f_{ii}^{(n)}$ as the probability that the first return to state i occurs at time n. By definition we say $f_{ij}^{(0)} = f_{ii}^{(0)} = 0$.

Remark.   G.C.D. $\{n : p_{jj}^{(n)} > 0\} = $ G.C.D. $\{n : f_{jj}^{(n)} > 0\}$ (Exercise 8).

Definition II.1.6.   *For fixed states* i *and* j, *let* $f_{ij}^* = \sum_{n=1}^{\infty} f_{ij}^{(n)}$. *The symbol* f_{ij}^*
represents the probability of ever visiting state j *from state* i. *If* $i = j$, *we let*
$f_{ii}^* = \sum_{n=1}^{\infty} f_{ii}^{(n)}$ *denote the probability of ultimately returning to state* i.

Definition II.1.7.   *A state* j *is said to be persistent if* $f_{jj}^* = 1$. *If* $f_{jj}^* < 1$, *then* j *is
called* transient.

The following example will illustrate these concepts.

Example II.1.5.   Let $\{X_n\}$ be a Markov chain with state space, $\{1, 2, 3, 4\}$
and transition matrix

$$P = \begin{bmatrix} \frac{1}{2} & \frac{1}{2} & 0 & 0 \\ 1 & 0 & 0 & 0 \\ 0 & \frac{1}{3} & \frac{2}{3} & 0 \\ \frac{1}{2} & 0 & \frac{1}{2} & 0 \end{bmatrix}.$$

In this case $f_{44}^{(n)} = 0$ for all n so state 4 is surely transient. For state 3 we
have that $f_{33}^{(1)} = \frac{2}{3}$, $f_{33}^{(n)} = 0$ for $n \geqslant 2$ so state 3 is transient. States 1 and 2 are
persistent as the following arguments show:

$$f_{11}^* = f_{11}^{(1)} + f_{11}^{(2)} = \tfrac{1}{2} + \tfrac{1}{2} = 1$$

$$f_{22}^* = f_{22}^{(1)} + f_{22}^{(2)} + \cdots + f_{22}^{(k)} + \cdots = 0 + \tfrac{1}{2} + \tfrac{1}{4} + \tfrac{1}{8} + \cdots = 1.$$

In the case where state j is persistent, the numbers $f_{jj}^{(n)}$ form a probability
distribution on the times of the first return. That is $f_{jj}^{(n)} \geqslant 0$ for all n and
$\sum_{n=1}^{\infty} f_{jj}^{(n)} = 1$. Hence it is reasonable to consider the expected time until the
first return.

Definition II.1.8.   *If* $f_{jj}^* = 1$, *define the expected return time to state* j *as*
$\mu_j = \sum_{n=1}^{\infty} n f_{jj}^{(n)}$.

For reasons that will become clear later, a persistent state with an
infinite expected return time is called *null persistent*. If the expected return
time is finite, the state is called *positive persistent*.

Example II.1.6.   Consider the Markov chain of Example II.1.5. States 1
and 2 were shown to be persistent so the expected return time to these

states can be calculated:

$$\mu_1 = \sum_{k=1}^{\infty} k f_{11}^{(k)} = 1 \cdot \tfrac{1}{2} + 2 \cdot \tfrac{1}{2} = \tfrac{3}{2}$$

$$\mu_2 = \sum_{k=2}^{\infty} k f_{22}^{(k)} = \sum_{k=2}^{\infty} \frac{k}{2^{k-1}} = 3.$$

Hence both of these states are positive persistent.

Example II.1.7. Let $\{X_n\}$ be a Markov chain with state space, $S = \{1,2,3,\dots\}$. Let the transition matrix be given by

$$P = \begin{bmatrix}
\frac{1}{2} & \frac{1}{2} & 0 & 0 & 0 & \cdot & \cdot & \cdot \\
\frac{1}{2} & 0 & \frac{1}{2} & 0 & 0 & \cdot & \cdot & \cdot \\
\frac{1}{2} & 0 & 0 & \frac{1}{2} & 0 & \cdot & \cdot & \cdot \\
\frac{1}{2} & 0 & 0 & 0 & \frac{1}{2} & \cdot & \cdot & \cdot \\
\cdot & \cdot & \cdot & \cdot & \cdot & \cdot & \cdot & \cdot
\end{bmatrix}.$$

Now $f_{11}^{(1)} = \tfrac{1}{2}$, $f_{11}^{(2)} = \tfrac{1}{4}$, $f_{11}^{(3)} = \tfrac{1}{8}$, $f_{11}^{(4)} = \tfrac{1}{16}$ and in general $f_{11}^{(n)} = 1/2^n$ so $f_{11}^* = 1$. However it is not so easy to find f_{ii}^* for $i \geq 2$ since various loops among the states must be considered. In view of this, a second characterization of persistence would be helpful. This second characterization will be given in Section 2.2 in terms of the $\{ p_{ii}^{(k)} \}$, $k = 0,1,2,\dots$.

EXERCISES

1. Find the G.C.D. of S where
   (a) $S = \{0,8,12,16\}$
   (b) $S = \{3,9,15,25\}$
   (c) $S = \{16,24,28,30\}$.

2. Let $\{X_n\}$ be a Markov chain with transition matrix given by

(a) $P = \begin{bmatrix}
.5 & 0 & 0 & .5 & 0 & 0 \\
0 & .4 & 0 & 0 & .6 & 0 \\
0 & .2 & .3 & 0 & 0 & .5 \\
0 & 0 & 0 & .3 & 0 & .7 \\
0 & 1 & 0 & 0 & 0 & 0 \\
.2 & 0 & 0 & 0 & 0 & .8
\end{bmatrix}$

(b) $P = \begin{bmatrix}
.5 & 0 & .1 & .4 & 0 & 0 \\
0 & 1 & 0 & 0 & 0 & 0 \\
0 & .2 & .3 & 0 & 0 & .5 \\
0 & 0 & 0 & .3 & 0 & .7 \\
0 & 1 & 0 & 0 & 0 & 0 \\
.2 & 0 & 0 & 0 & 0 & .8
\end{bmatrix}.$

For each chain, find all of the closed subsets of S which themselves have no proper closed subsets.

3. Let $\{X_n\}$ be a discrete-time stationary Markov chain with state space, $S = \{1, 2, 3, 4\}$. Let the transition matrix be given by

$$P = \begin{bmatrix} 0 & 0 & \frac{1}{2} & \frac{1}{2} \\ 0 & 1 & 0 & 0 \\ 0 & 0 & 0 & 1 \\ 1 & 0 & 0 & 0 \end{bmatrix}.$$

Classify the states according to persistency and periodicity. For the persistent states find the expected return time.

4. Do problem 3 using the stochastic matrix

$$P = \begin{bmatrix} 0 & 0 & 1 & 0 \\ 1 & 0 & 0 & 0 \\ 0 & \frac{1}{2} & \frac{1}{2} & 0 \\ \frac{1}{3} & 0 & 0 & \frac{2}{3} \end{bmatrix}.$$

5. Let j be a transient state in some state space, S. Define $C_j = \{k : p_{jk}^{(m)} > 0 \text{ for some } m\}$. Is C_j an irreducible closed set?

6. Determine whether or not state 2 is persistent or transient by finding f_{22}^* for the Markov chain defined in (a) Example II.1.4, (b) Example II.1.7.

7. Prove Proposition II.1.1.

8. Show that G.C.D. $\{n : p_{ii}^{(n)} > 0\} = $ G.C.D. $\{n : f_{ii}^{(n)} > 0\}$.

SECTION 2   **CLASSIFICATION OF STATES**

At the conclusion of Section II.1 it was stated that a second characterization of persistence would be given using the $p_{ii}^{(n)}$'s. In order to do this we will use several of the mathematical results in Section I.4.

The first step is to establish a relationship between the $p_{ij}^{(n)}$'s and the $f_{ij}^{(n)}$'s. An equation showing this relationship is

$$p_{ij}^{(n)} = \sum_{k=0}^{n} f_{ij}^{(k)} p_{jj}^{(n-k)} \qquad \text{for } n \geq 1. \qquad (2.1)$$

(Equation 2.1 fails to hold for $n = 0$ when $i = j$ since in this case the left-hand side is one while the right-hand side is zero; see Definitions I.3.4 and II.1.5.) This equation says that if the chain goes from state i to state j in n steps, the various paths having this property can be partitioned into mutually exclusive sets according to when they make their first visit to state j. (If $i = j$ we speak of the first return rather than first visit). To see how (2.1) is derived, note that

$$P[X_n = j | X_0 = i] = \sum_{k=1}^{n} P[X_n = j, X_k = j, X_{k-1} \neq j, \ldots X_1 \neq j | X_0 = i].$$

Now use the fact that

$$P[X_n = j, X_k = j, X_{k-1} \neq j, \ldots X_1 \neq j | X_0 = i]$$

$$= P[X_k = j, X_{k-1} \neq j, \ldots X_1 \neq j | X_0 = i] \cdot P[X_n = j | X_k = j,$$

$$X_{k-1} \neq j, \ldots X_1 \neq j, X_0 = i]$$

$$= f_{ij}^{(k)} p_{jj}^{(n-k)}.$$

[Note that (2.1) could also be written as $p_{ij}^{(n)} = \sum_{k=0}^{n} f_{ij}^{(n-k)} p_{jj}^{(k)}$ (Exercise 5). There are times when this representation is more useful; see the proof of Theorem III.5.1.]

Remark II.2.1. At this time the reader should compare (2.1) with the Chapman–Kolmogorov identity, in Chapter I. Both of these yield ways of expressing $p_{ij}^{(n)}$ in terms of sums. The distinctions between these sums should be considered and clearly understood by the reader.

Let i and j denote states in the state space, S, of a Markov chain. Let $\{p_{ij}^{(n)}\}_{n=0}^{\infty}$ denote the sequence of n step probabilities of visits to state j from state i with corresponding generating function, $\mathcal{P}_{ij}(s) = \sum_{n=0}^{\infty} p_{ij}^{(n)} s^n$. Let $\{f_{ij}^{(n)}\}_{n=0}^{\infty}$ denote the sequence of n step probabilities of first visits to state j from state i with corresponding generating function, $\mathcal{F}_{ij}(s) = \sum_{n=0}^{\infty} f_{ij}^{(n)} s^n$. The following lemma establishes a correspondence between $\mathcal{F}_{ij}(s)$ and $\mathcal{P}_{ij}(s)$.

Lemma II.2.1. For any states i and j in the state space of a Markov chain,

$$\mathcal{P}_{ii}(s) = \frac{1}{1 - \mathcal{F}_{ii}(s)} \qquad \text{for } s \in (-1, 1),$$

and

$$\mathcal{P}_{ij}(s) = \mathcal{F}_{ij}(s) \cdot \mathcal{P}_{jj}(s) \qquad \text{for } s \in (-1, 1) \qquad \text{when } i \neq j.$$

Proof. From (2.1) we get

$$p_{ii}^{(n)} = \sum_{k=0}^{n} f_{ii}^{(k)} p_{ii}^{n-k} \qquad \text{for } n \geqslant 1$$

so

$$\mathcal{P}_{ii}(s) = \sum_{n=0}^{\infty} p_{ii}^{(n)} s^n = 1 + \sum_{n=1}^{\infty} p_{ii}^{(n)} s^n$$

$$= 1 + \sum_{n=1}^{\infty} \left(\sum_{k=0}^{n} f_{ii}^{(k)} p_{ii}^{(n-k)} \right) s^n.$$

The next step is to interchange the order of summation using Fubini's theorem. Before making the interchange, the applicability of the theorem will be discussed. If $s \in (0, 1)$ then all terms are non-negative so the interchange is justified. However, it was assumed that $s \in (-1, 1)$ so the absolute convergence of the series must be shown. It is easy to see that

$$\sum_{n=1}^{\infty} p_{ii}^{(n)} s^n \leqslant \sum_{n=1}^{\infty} p_{ii}^{(n)} |s|^n < \infty \qquad \text{for } s \in (-1, 1),$$

so Fubini's theorem can be used. Now if we define $p_{ii}^{(m)} = 0$ for $m < 0$, the double series assumes the form of Fubini's theorem and we get

$$\mathcal{P}_{ii}(s) = 1 + \sum_{n=1}^{\infty} \sum_{k=0}^{\infty} f_{ii}^{(k)} p_{ii}^{(n-k)} s^n$$

$$= 1 + \sum_{k=0}^{\infty} f_{ii}^{(k)} s^k \sum_{n=1}^{\infty} p_{ii}^{(n-k)} s^{n-k}$$

$$= 1 + \sum_{k=0}^{\infty} f_{ii}^{(k)} s^k \sum_{n=k}^{\infty} p_{ii}^{(n-k)} s^{n-k}$$

$$= 1 + \sum_{k=0}^{\infty} f_{ii}^{(k)} s^k \mathcal{P}_{ii}(s)$$

$$= 1 + \mathcal{F}_{ii}(s) \mathcal{P}_{ii}(s).$$

Simplifying this expression we get $\mathcal{P}_{ii}(s) = 1/[1 - \mathcal{F}_{ii}(s)]$ for $-1 < s < 1$.
     The proof of the second part of Lemma II.2.1 is left as an exercise.   ▲

     Lemma II.2.1 and Abel's theorem combine to give the characterization of persistence mentioned at the beginning of this section.

Theorem II.2.1.

State i is persistent if and only if $\sum_{n=0}^{\infty} p_{ii}^{(n)} = \infty$.

Proof. By Lemma II.2.1 we have $\mathcal{P}_{ii}(s) = 1/[1 - \mathcal{F}_{ii}(s)]$. Assume state i is persistent, which means $\sum_{n=0}^{\infty} f_{ii}^{(n)} = 1$. By Abel's theorem this implies that $\lim_{s \to 1^-} \mathcal{F}_{ii}(s) = 1$ so

$$\lim_{s \to 1^-} \mathcal{P}_{ii}(s) = \lim_{s \to 1^-} \frac{1}{1 - \mathcal{F}_{ii}(s)} = \infty.$$

But since the coefficients in the series $\mathcal{P}_{ii}(s)$ are non-negative, Abel's theorem applies again to yield $\sum_{n=0}^{\infty} p_{ii}^{(n)} = \infty$.

Conversely if $\sum_{n=0}^{\infty} f_{ii}^{(n)} < 1$, then by Abel's theorem $\lim_{s \to 1^-} \mathcal{F}_{ii}(s) < 1$. Hence by Lemma II.2.1 we get

$$\lim_{s \to 1^-} \mathcal{P}_{ii}(s) = \frac{1}{1 - \lim_{s \to 1^-} \mathcal{F}_{ii}(s)} < \infty.$$

Since $p_{ii}^{(n)} \geqslant 0$, we apply Abel's theorem once more to get $\sum_{n=0}^{\infty} p_{ii}^{(n)} < \infty$. This completes the proof. ▲

We are now in the position of being able to classify each state of a Markov chain according to persistency and periodicity. However the conditions

$$\sum_{n=0}^{\infty} p_{ii}^{(n)} = \infty \quad \text{or} \quad \sum_{n=0}^{\infty} f_{ii}^{(n)} = 1$$

and

$$\text{G.C.D.} \{ n : p_{ii}^{(n)} > 0 \} = 1 \quad \text{or} \quad \text{G.C.D.} \{ n : f_{ii}^{(n)} > 0 \} = 1$$

are not always easy to check. In particular, classifying all the states of a Markov chain with, say, 25 states would be very tedious. The following theorem makes this job much easier since it says that states that intercommunicate are the same type with regard to persistency and periodicity.

Theorem II.2.2.

If states i and j intercommunicate, they are of the same type.

Proof. Let i and j be arbitrary states in S that intercommunicate. This means that there exist integers m and n such that $p_{ij}^{(m)} = \alpha > 0$ and $p_{ji}^{(n)} = \beta > 0$. Hence for all integers r we have $p_{jj}^{(r+m+n)} \geqslant p_{ji}^{(n)} p_{ii}^{(r)} p_{ij}^{(m)} = \alpha \beta p_{ii}^{(r)}$. The inequality arises since there are fewer sample paths going from j to j in $r + m + n$ steps with scheduled stops at state i at times n and $n + r$ than there are unrestricted paths going from j to j in $r+m+n$ steps. Now if j is transient, the sum of the left-hand side over r is finite so $\sum_{r=0}^{\infty} p_{ii}^{(r)} < \infty$ and hence i is transient.

If j is persistent, consider $p_{ii}^{(r+m+n)} \geqslant p_{ij}^{(m)} p_{jj}^{(r)} p_{ji}^{(n)} = \alpha \cdot \beta \cdot p_{jj}^{(r)}$. If $\sum_{r=0}^{\infty} p_{jj}^{(r)} = \infty$, then

$$\infty = \alpha\beta \sum_{r=0}^{\infty} p_{jj}^{(r)} \leqslant \sum_{r=0}^{\infty} p_{ii}^{(r+m+n)} \leqslant \sum_{k=0}^{\infty} p_{ii}^{(k)}$$

and hence i is persistent. It is left to the reader to show that if j is null persistent, then i is null persistent and if j is positive persistent then i is positive persistent (Exercise 13, Section III.2).

Again let i and j be arbitrary states in S that intercommunicate and let the period of state i be d. Then $p_{ii}^{(n)} = 0$ unless n is a multiple of d. Now $p_{ii}^{(r+m+n)} \geqslant p_{ij}^{(m)} p_{jj}^{(r)} p_{ji}^{(n)} = \alpha\beta p_{jj}^{(r)}$ as before, so for $r=0$ we get $p_{ii}^{(m+n)} > 0$. Hence $m+n$ is a multiple of d. Now $p_{ii}^{(r+m+n)}$ will be zero unless r is a multiple of d, so $p_{jj}^{(r)}$ will be zero unless r is a multiple of d. Therefore the period of j is a multiple of d. Repeating the above argument using $p_{jj}^{(r+m+n)} \geqslant \alpha\beta p_{ii}^{(r)}$ and using the fact that the period of j is $k \cdot d$, it follows that the period of i is a multiple of $k \cdot d$. But the period of i is d so k must equal one. Hence i and j have the same period and since they were arbitrary states, the proof is complete. ▲

Corollary II.2.1. All states belonging to an irreducible closed subset of S are of the same type.

Proof. First note that an irreducible closed subset of S can be considered the entire state space for a Markov chain that starts within the closed set. Hence an irreducible closed subset of S can be viewed as a state space for an irreducible Markov chain. Using Theorem II.1.1 and the above theorem the corollary follows. ▲

Let us again consider Example II.1.7. In that example we were able to show that state 1 is persistent. We also indicated that the other states could not easily be classified because it is a nontrivial task to find the sequence $\{f_{ii}^{(n)}\}_{n=1}^{\infty}$ for $i \geqslant 2$. We used this example to illustrate the need for an alternative (and hopefully simpler) means for showing persistency. An alternative means was provided in Theorem II.2.1. However, if we were to try using this theorem for showing persistency of the states $i \geqslant 2$, we would find that the $p_{ii}^{(n)}$ values are no easier to find than the $f_{ii}^{(n)}$ values. Does this mean that we have been guilty of "false advertising"? Not at all! The results of Theorem II.2.1 were crucial to the proof of Theorem II.2.2, which does provide for Example II.1.7 a simple means for determining persistency. In particular, since all the states intercommunicate and since state 1 is persistent, it follows that all of the states are persistent.

In classifying the states of a Markov chain, the simplest cases are those like Example II.1.7 where the Markov chain is irreducible and hence where there is only one nonempty irreducible closed subset of S, namely S itself. In this case all states are classified once any single state in S is classified.

Example II.2.1. Consider a Markov chain with transition matrix

$$P = \begin{bmatrix} \frac{1}{4} & \frac{1}{4} & \frac{1}{4} & \frac{1}{4} \\ 0 & 0 & 1 & 0 \\ 0 & 0 & 0 & 1 \\ 1 & 0 & 0 & 0 \end{bmatrix}.$$

It is easy to see that this chain is irreducible. Now it turns out that state 1 is the easiest to classify. In particular state 1 is certainly aperiodic since $p_{11}^{(1)} > 0$. [A positive value in the (i,i)th position of P always makes state i aperiodic.] The persistency of state 1 is also easy to show since $f_{11}^{(1)} = \frac{1}{4}$, $f_{11}^{(2)} = \frac{1}{4}$, $f_{11}^{(3)} = \frac{1}{4}$, $f_{11}^{(4)} = \frac{1}{4}$ so $f_{11}^* = 1$. The expected return time to state 1 is $\mu_1 = 1 \cdot \frac{1}{4} + 2 \cdot \frac{1}{4} + 3 \cdot \frac{1}{4} + 4 \cdot \frac{1}{4} = 10/4$. Hence state 1 is positive persistent. Theorem II.2.2 now implies that all the states are positive persistent aperiodic. If one decided to use state 2 to check for persistency, an infinite series for f_{22}^* would be obtained. In view of this the reader is advised to choose carefully the state in an irreducible class that he wishes to classify.

Unfortunately not all Markov chains are irreducible, so the next question to consider is that of identifying the irreducible closed subsets of S. Then, as shown above, all the states within one of these irreducible closed subsets are of the same type. The following theorem is helpful for finding irreducible closed subsets of S.

Theorem II.2.3.

Let j be a persistent state in some state space, S. There exists a unique irreducible closed set, $\mathcal{C}$, containing j and such that for every pair of states i and k in $\mathcal{C}$, $f_{ik}^* = 1$ and $f_{ki}^* = 1$. $\mathcal{C}$ is often referred to as the irreducible closed set generated by state j.

Proof. $\mathcal{C}$ is defined to be the set of all states that can be reached from j. $\mathcal{C}$ is clearly closed since by construction states outside of $\mathcal{C}$ are states that cannot be reached. For $i \in \mathcal{C}$ we know that the probability of going from j to i is positive. Let $0 < \alpha = P$ [the chain eventually gets to i|the chain starts at j]. Since the probability of *not* returning to j from i is $1 - f_{ij}^*$, the probability of not returning to j from j is greater than or equal to α [$1 - f_{ij}^*$].

But since j is persistent, the probability of not returning to j from j is zero so $f_{ij}^* = 1$. Similarly if k is some other state in $\mathcal{C}, f_{kj}^* = 1$.

Now the probability of going from i to k is positive since from i the chain can reach j and from j it can reach k. Similarly, it is possible to go from k to i. Hence the states within $\mathcal{C}$ intercommunicate so $\mathcal{C}$ is an irreducible state space. Therefore all states within $\mathcal{C}$ are persistent and any state in $\mathcal{C}$ can play the role of the state j above. Therefore $f_{ki}^* = 1$ and $f_{ik}^* = 1$ by letting i and then k play the role of j above. ▲

The problem of classification of states now takes the following form. If it is known that the chain is irreducible, any state can be used to classify all the states. If the chain is not irreducible, the first step is to identify the various irreducible closed subsets of S. One way to find a closed subset is to start with a persistent state, j, and generate the irreducible closed subset $\mathcal{C}_j$. All states in $\mathcal{C}_j$ are persistent and they have the same period. (Possibly some state in $\mathcal{C}_j$ other than j could be used to determine the periodicity.)

The first step in this method of classifying states is the difficult one, namely finding a persistent state. One approach to this would be to use either of the characterizations $\sum_{n=0}^{\infty} p_{ii}^{(n)} = \infty$ or $\sum_{n=0}^{\infty} f_{ii}^{(n)} = 1$. Alternatively a diagram such as Figure II.1.1 might be used. For a persistent state, every path that leaves the state must eventually return. Other methods for finding persistent states will be considered in Chapter VI. The following example demonstrates the ideas mentioned above.

Example II.2.2. Let $\{X_n\}$ be a Markov chain with the transition matrix

$$P = \begin{bmatrix} 0 & 0 & 1 & 0 & 0 & 0 \\ 0 & 0 & 0 & 0 & 0 & 1 \\ 0 & 0 & 0 & 0 & 1 & 0 \\ \frac{1}{3} & \frac{1}{3} & 0 & \frac{1}{3} & 0 & 0 \\ 1 & 0 & 0 & 0 & 0 & 0 \\ 0 & \frac{1}{2} & 0 & 0 & 0 & \frac{1}{2} \end{bmatrix}.$$

Let $S = \{1, 2, 3, 4, 5, 6\}$ with the usual correspondence to P. State 1 is persistent since $f_{11}^{(1)} = 0$, $f_{11}^{(2)} = 0$, $f_{11}^{(3)} = 1$. The irreducible closed set generated by state 1 is $\mathcal{C}_1 = \{1, 3, 5\}$. The period of these three states is three since G.C.D. $\{n : p_{11}^{(n)} > 0\} = 3$. A second persistent state is state 6 since $f_{66}^{(1)} = \frac{1}{2}$ and $f_{66}^{(2)} = \frac{1}{2}$. The irreducible closed set generated by state 6 is $\mathcal{C}_6 = \{2, 6\}$. Since $p_{66}^{(1)} > 0$ these two states are aperiodic. For state 4 we have $f_{44}^{(1)} = \frac{1}{3}$ and $f_{44}^{(k)} = 0$ for $k \geqslant 2$ so state 4 is transient. For a transient state the period is usually not of interest so we will not calculate the period of state 4. This completes a classification of the states. (Note that since the state space is finite, all the persistent states are positive persistent. This fact will be verified in Section III.2.)

The diagram given in Figure II.2.1 might also be useful in the above classification. At a glance one can see that $\{1,3,5\}$ and $\{2,6\}$ form closed sets. At this point one of the states in each closed set must be classified. The remaining state is transient.

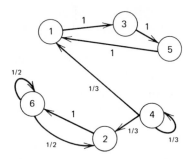

Figure II.2.1.   Digraph for the matrix of Example II.2.2.

In addition to simplifying the problem of classifying the states in S, the identification of the various irreducible closed sets serves another purpose. If the rows and columns of P are interchanged so that the states within these irreducible closed sets are grouped together, the new transition matrix assumes a *block form*. That is, in the above example, if the columns and rows of the transition matrix corresponded to states $1,3,5,2,6,4$ respectively, the matrix, P^*, would have the form

$$P^* = \begin{bmatrix} 0 & 1 & 0 & 0 & 0 & 0 \\ 0 & 0 & 1 & 0 & 0 & 0 \\ 1 & 0 & 0 & 0 & 0 & 0 \\ 0 & 0 & 0 & 1 & 0 & 0 \\ 0 & 0 & 0 & \frac{1}{2} & \frac{1}{2} & 0 \\ \frac{1}{3} & 0 & 0 & \frac{1}{3} & 0 & \frac{1}{3} \end{bmatrix}.$$

This is essentially the same transition matrix as P except that the rows and columns correspond to different states. P^* is said to be in block form since it can be partitioned into blocks as follows.

$$P^* = \begin{bmatrix} 0 & 1 & 0 & 0 & 0 & 0 \\ 0 & 0 & 1 & 0 & 0 & 0 \\ 1 & 0 & 0 & 0 & 0 & 0 \\ 0 & 0 & 0 & 1 & 0 & 0 \\ 0 & 0 & 0 & \frac{1}{2} & \frac{1}{2} & 0 \\ \frac{1}{3} & 0 & 0 & \frac{1}{3} & 0 & \frac{1}{3} \end{bmatrix}.$$

There is a block corresponding to each of the irreducible closed subsets of S. Note that each of these blocks is a stochastic matrix itself. The transitions among the transient states are placed in the lower right-hand corner of P^*.

In general, a stochastic matrix written in block form is as follows:

$$
P^* = \begin{bmatrix}
P_1 & 0 & 0 & . & . & . & 0 & 0 \\
0 & P_2 & 0 & . & . & . & 0 & 0 \\
0 & 0 & P_3 & . & . & . & 0 & 0 \\
. & . & . & . & . & . & . & . \\
. & . & . & . & . & . & . & . \\
0 & 0 & 0 & . & . & . & P_n & 0 \\
R_1 & R_2 & R_3 & . & . & . & R_n & Q
\end{bmatrix}.
$$

Each P_i denotes the stochastic matrix corresponding to transitions within the irreducible closed subset $\mathcal{C}_i$. The matrix Q corresponds to transitions among the transient states T. The matrix Q is not stochastic since $\sum_{j \in T} q_{ij}$ $\leqslant 1$ for each $i \in T$. (A matrix of non-negative elements with the property that the row sums are less than or equal to one will be called *substochastic*; see Definition III.3.1.) The matrices $R_1, R_2, \ldots, R_n$ correspond to transitions from the transient states into the persistent states. The R_i matrices are not square matrices in general so they do not have an interpretation as a transition matrix from a state space into itself.

The following example shows how an analysis of a chain is much easier if P is written in block form.

Example II.2.3. Let P be a stochastic matrix for a Markov chain with state space, $S = \{1, 2, \ldots, 10\}$.

$$
P = \begin{bmatrix}
\frac{1}{3} & 0 & 0 & 0 & \frac{2}{3} & 0 & 0 & 0 & 0 & 0 \\
0 & \frac{1}{2} & 0 & \frac{1}{2} & 0 & 0 & 0 & 0 & 0 & 0 \\
0 & 0 & \frac{1}{2} & 0 & \frac{1}{2} & 0 & 0 & 0 & 0 & 0 \\
0 & 0 & 0 & 0 & 0 & 0 & 0 & 1 & 0 & 0 \\
1 & 0 & 0 & 0 & 0 & 0 & 0 & 0 & 0 & 0 \\
0 & 0 & 0 & 0 & 0 & 1 & 0 & 0 & 0 & 0 \\
\frac{1}{4} & 0 & \frac{1}{4} & 0 & \frac{1}{4} & \frac{1}{4} & 0 & 0 & 0 & 0 \\
0 & 0 & 0 & 0 & 0 & 0 & 0 & 0 & 1 & 0 \\
0 & 1 & 0 & 0 & 0 & 0 & 0 & 0 & 0 & 0 \\
\frac{1}{10} & \frac{1}{10} & \frac{1}{10} & \frac{1}{10} & \frac{1}{10} & \frac{1}{10} & \frac{1}{10} & \frac{1}{10} & \frac{1}{10} & \frac{1}{10}
\end{bmatrix}.
$$

If one of the above methods is used to reduce P to block form we get the following matrix, P^*. The rows and columns of P^* correspond to the states

1, 5, 2, 4, 8, 9, 6, 3, 7, 10, respectively.

$$P^* = \begin{pmatrix} \frac{1}{3} & \frac{2}{3} & 0 & 0 & 0 & 0 & 0 & 0 & 0 & 0 \\ 1 & 0 & 0 & 0 & 0 & 0 & 0 & 0 & 0 & 0 \\ 0 & 0 & \frac{1}{2} & \frac{1}{2} & 0 & 0 & 0 & 0 & 0 & 0 \\ 0 & 0 & 0 & 0 & 1 & 0 & 0 & 0 & 0 & 0 \\ 0 & 0 & 0 & 0 & 0 & 1 & 0 & 0 & 0 & 0 \\ 0 & 0 & 1 & 0 & 0 & 0 & 0 & 0 & 0 & 0 \\ 0 & 0 & 0 & 0 & 0 & 0 & 1 & 0 & 0 & 0 \\ 0 & \frac{1}{2} & 0 & 0 & 0 & 0 & 0 & \frac{1}{2} & 0 & 0 \\ \frac{1}{4} & \frac{1}{4} & 0 & 0 & 0 & 0 & \frac{1}{4} & \frac{1}{4} & 0 & 0 \\ \frac{1}{10} & \frac{1}{10} & \frac{1}{10} & \frac{1}{10} & \frac{1}{10} & \frac{1}{10} & \frac{1}{10} & \frac{1}{10} & \frac{1}{10} & \frac{1}{10} \end{pmatrix}.$$

From P^* it is easy to see that states 3, 7, and 10 are transient. The classification of the states within each closed set is easy since these sets are small. In fact all the remaining states are positive persistent and aperiodic.

Using the techniques of classifying states and using the mathematical results of Chapter I we return in Chapter III to the main question of determining when a long run distribution exists and finding it when it does.

EXERCISES

1. Let $\{X_n\}$ be a discrete-time stationary Markov chain with transition matrix P as given below and state space, $S = \{1, 2, \ldots, 10\}$.

    i.   Rewrite P into block form.

   ii.   Classify the states in S.

$$P = \begin{pmatrix} 0 & 0 & 0 & \frac{1}{2} & 0 & 0 & \frac{1}{2} & 0 & 0 & 0 \\ 0 & 0 & 0 & 0 & 0 & 0 & 0 & 0 & 0 & 1 \\ \frac{1}{3} & 0 & \frac{1}{3} & 0 & 0 & 0 & 0 & \frac{1}{3} & 0 & 0 \\ \frac{1}{4} & 0 & 0 & 0 & 0 & 0 & \frac{3}{4} & 0 & 0 & 0 \\ 0 & 0 & 0 & 0 & 0 & \frac{1}{2} & 0 & \frac{1}{2} & 0 & 0 \\ 0 & 0 & 0 & 0 & 1 & 0 & 0 & 0 & 0 & 0 \\ \frac{1}{5} & 0 & 0 & \frac{4}{5} & 0 & 0 & 0 & 0 & 0 & 0 \\ 0 & 0 & 0 & 0 & 1 & 0 & 0 & 0 & 0 & 0 \\ \frac{1}{4} & 0 & \frac{1}{4} & 0 & 0 & 0 & \frac{1}{4} & 0 & \frac{1}{4} & 0 \\ 0 & 1 & 0 & 0 & 0 & 0 & 0 & 0 & 0 & 0 \end{pmatrix}.$$

2. Write a transition matrix for a Markov chain that has state space $S = \{1, 2, 3, 4, 5, 6, 7, 8, 9\}$ where states 1, 4, and 7 form a closed set of period 2, states 2, 3, and 5 form an aperiodic closed subset, state 9 is absorbing, and the remaining states are transient.

3. Let i be a state of a Markov chain for which $f_{ii}^{(n)} = n/2^{n+1}$ for $n = 0, 1, 2, \ldots$.

   i. Find $\mathcal{P}_{ii}(s)$.

   ii. Show that state i is persistent.

   iii. Find the mean recurrence time to state i.

4. Prove or give a counterexample to the following:

   (a) If state i is transient and $p_{ij}^{(m)} > 0$ for some m, then j is transient.

   (b) If state j is transient and $p_{ij}^{(m)} > 0$ for some m, then i is transient.

   (c) If the state space, S, of an irreducible Markov chain has n states, then the period of every state in S divides n.

   (d) If P_1 and P_2 are $n \times n$ stochastic matrices corresponding to irreducible aperiodic Markov chains, then $P_1 P_2$ is a stochastic matrix corresponding to an irreducible aperiodic Markov chain.

5. Show that $p_{ij}^{(n)} = \sum_{k=0}^{n} f_{ij}^{(n-k)} p_{jj}^{(k)}$ for $n \geqslant 1$.

*6. Prove that for any two distinct states i and j in the state space, S,
   $$\mathcal{P}_{ij}(s) = \mathcal{F}_{ij}(s) \cdot \mathcal{P}_{jj}(s).$$

*7. (a) Show that if the jth column of P^m has no zeros then the jth column of P^{m+1} has no zeros.

   (b) Show that if the jth column of P^m has no zeros for some integer m, then state j is aperiodic.

8. Let P_1 and P_2 be stochastic matrices. Let R be defined by $R = \lambda P_1 + (1 - \lambda) P_2$ where $0 < \lambda < 1$. Show that if P_1 is the transition matrix of an irreducible aperiodic Markov chain, then so is R.

CHAPTER III

The Classical Approach to Markov Chains

SECTION 1 THE RENEWAL THEOREM

One of the main questions considered in this book is the question of what happens to $p_{ij}^{(n)}$ as $n \to \infty$. In particular, does this sequence converge and if it does, is the limit independent of i? Conditions will be given later that do guarantee such behavior, but before stating the conditions we will prove a form of the renewal theorem that is the key result for establishing the convergence of $p_{ij}^{(n)}$.

We begin with a lemma from algebra that is needed in the proof of the renewal theorem.

Lemma III.1.1. Let $a_1, a_2, \ldots, a_n$ be positive integers with greatest common divisor (G.C.D.) equal to 1. There exists an integer N such that if $m \geqslant N$, then m can be expressed as a positive linear combination of the a_i's. That is $m = \sum_{i=1}^{n} b_i a_i$ where the b_i's are *non-negative integers*.

Proof. It is well known that if 1 is the G.C.D. of $\{a_1, a_2, \ldots, a_n\}$, then there exist integers k_i (not necessarily non-negative) such that $1 = \sum_{i=1}^{n} k_i a_i$. [This result is contained in most undergraduate texts on algebra. See, for example, Birkhoff and MacLane (1953).] Let $s = \sum_{i=1}^{n} a_i$. Every positive integer m can be uniquely written as $m = qs + r$ where q and r are integers and $0 \leqslant r < s$. That is, divide s into m and get a quotient, q, and a remainder r. Now

$$m = q \sum_{i=1}^{n} a_i + r \sum_{i=1}^{n} k_i a_i = \sum_{i=1}^{n} (q + rk_i) a_i,$$

so simply choose m so large that $q \geqslant s|k_i|$ for all $i = 1, 2, \ldots, n$. Then all the coefficients, $q + rk_i$, will be non-negative. ▲

It should be mentioned that there are many forms of the renewal theorem. The forms differ in the degree of generality in which the theorem is stated and proved. Since we have a specific purpose in mind for the renewal theorem, we will give a form that is general enough to cover our needs.

Theorem III.1.1 (Renewal Theorem).

Let $\{a_k\}$, $\{b_k\}$, and $\{u_k\}$ be sequences of non-negative real numbers with $\sum_{k=0}^{\infty} a_k = 1$, $\sum_{k=0}^{\infty} b_k < \infty$, and with the $\{u_k\}$ sequence bounded. Assume the G.C.D.$\{k: a_k > 0\} = 1$ and assume the renewal equation $u_n - \sum_{k=0}^{n} a_{n-k} u_k = b_n$ holds for all $n = 0, 1, 2, 3, \ldots$. Then $\lim_{n\to\infty} u_n$ exists. In fact

$$\lim_{n\to\infty} u_n = \frac{\sum_{k=0}^{\infty} b_k}{\sum_{k=1}^{\infty} k a_k} \quad \text{if} \quad \sum_{k=1}^{\infty} k a_k < \infty$$

and

$$\lim_{n\to\infty} u_n = 0 \quad \text{if} \quad \sum_{k=1}^{\infty} k a_k = \infty.$$

Proof. The assumption that G.C.D. $\{k: a_k > 0\} = 1$ is very important and the theorem will be applied using this degree of generality. The notational difficulties that this condition introduces can be seen in the previous lemma. One can eliminate all of this difficulty by assuming $a_1 > 0$, but this not only improves notation but over simplifies the method of proof. As a compromise between these two extremes, the theorem will be proved assuming $a_1 = 0$, $a_2 > 0$, $a_3 > 0$.

Let $\lambda = \overline{\lim}_{n\to\infty} u_n$ and note that λ is finite by assumption. Choose a subsequence n_j so that $\lim_{j\to\infty} u_{n_j} = \lambda$. We will show that $\lim_{j\to\infty} u_{n_j-2} = \lambda$ using the condition that $a_2 > 0$. Assume it is false that $\lim_{j\to\infty} u_{n_j-2} = \lambda$. Then there exists $\lambda' < \lambda$ such that $u_{n_j-2} < \lambda'$ for infinitely many j. Let $\epsilon = a_2(\lambda - \lambda')/4$ and let $M = \sup_n u_n$. Choose N such that $\sum_{k=N+1}^{\infty} a_k \leqslant \epsilon/M$. Now, choose j so large that $n_j \geqslant N$, $u_{n_j} > \lambda - \epsilon$, $u_{n_j-2} < \lambda' < \lambda$, $b_{n_j} < \epsilon$, and $u_n < \lambda + \epsilon$, for all $n \geqslant n_j - N$. To see that this is possible first note that M and ϵ are fixed. Hence using the fact that $\sum_{k=0}^{\infty} a_k = 1$, there exists N such that $\sum_{k=N+1}^{\infty} a_k < \epsilon/M$. Fix such an N. Since $u_{n_j} \to \lambda$ as $j \to \infty$, there exists J_1 such that $\lambda - \epsilon < u_{n_j} < \lambda + \epsilon$ for all $j \geqslant J_1$. In fact since λ is the lim sup of the original sequence, $\{u_n\}$, all the terms of $\{u_n\}$ stay below $\lambda + \epsilon$ from some point on. That is, there exists J_2 such that $u_{n_j-N} < \lambda + \epsilon$ for all $j \geqslant J_2$. Also since $\sum_{k=0}^{\infty} b_k < \infty$, there exists J_3 such that $b_{n_j} < \epsilon$ for all $j \geq J_3$. Let $J = \max(J_1, J_2, J_3)$. For infinitely many j we have $u_{n_j - 2} < \lambda'$ so fix such a j^* that is larger than J. For this j^* the above conditions are satisfied.

Now

$$u_{n_{j^*}} = \sum_{k=0}^{n_{j^*}} a_k u_{n_{j^*} - k} + b_{n_{j^*}} < \sum_{k=0}^{N} a_k u_{n_{j^*} - k} + \sum_{k=N+1}^{n_{j^*}} a_k u_{n_{j^*} - k} + \epsilon$$

$$< \sum_{k=0}^{N} a_k u_{n_{j^*} - k} + M \sum_{k=N+1}^{\infty} a_k + \epsilon$$

$$< \sum_{k=0}^{N} a_k u_{n_{j^*} - k} + 2\epsilon < (a_0 + a_1 + a_3 + \cdots + a_N)(\lambda + \epsilon) + a_2 \lambda' + 2\epsilon$$

$$\leq (1 - a_2)(\lambda + \epsilon) + a_2 \lambda' + 2\epsilon < \lambda - (\lambda - \lambda')a_2 + 3\epsilon = \lambda - \epsilon.$$

This contradicts a previous inequality so it must be true that $\lim_{j \to \infty} u_{n_j - 2} = \lambda$. Using the fact that $a_3 > 0$ it is possible to show in exactly the same way that $\lim_{j \to \infty} u_{n_j - 3} = \lambda$. Now starting with $\lim_{j \to \infty} u_{n_j - 3} = \lambda$ and repeating the above argument we get $\lim_{j \to \infty} u_{n_j - 3 - 2} = \lambda$. Continuing this procedure we get $\lim_{j \to \infty} u_{n_j - 2\alpha - 3\beta} = \lambda$ for all positive integers α and β. By Lemma III.1.1 we know there exists an integer N^* such that $\lim_{j \to \infty} u_{n_j - m} = \lambda$ for $m > N^*$. (In fact, if $a_2 > 0$ and $a_3 > 0$ then $N^* = 1$.) In any case assume N^* is chosen to be the smallest integer with this property. We want to show that $\lim_{j \to \infty} u_{n_j - d} = \lambda$ for all positive integers, d, which means $N^* = 0$.

If $N^* > 0$ then by definition of N^*, $\lim_{j \to \infty} u_{n_j - N^*} \neq \lambda$. However, consider

$$u_{n_j - N^*} = \sum_{k=0}^{n_j - N^*} a_k u_{n_j - N^* - k} + b_{n_j - N^*}.$$

This can be rewritten as

$$(1 - a_0) u_{n_j - N^*} = \sum_{k=0}^{n_j - N^* - 1} a_{k+1} u_{n_j - N^* - (k+1)} + b_{n_j - N^*}$$

so since $\sum_{k=0}^{n_j - N^* - 1} a_{k+1} \to 1 - a_0$ as $j \to \infty$ and $u_{n_j - N^* - (k+1)} \to \lambda$ for all k as $j \to \infty$ and $b_{n_j - N} \to 0$ as $j \to \infty$, we get

$$\sum_{k=0}^{n_j - N^* - 1} a_{k+1} u_{n_j - N^* - (k+1)} \to \lambda(1 - a_0).$$

(See Theorem I.4.4.) Hence $(1 - a_0)\lim_{j \to \infty} u_{n_j - N^*} = (1 - a_0)\lambda$ and since $a_0 < 1$ we have $\lim_{j \to \infty} u_{n_j - N^*} = \lambda$. This contradiction implies that $\lim_{j \to \infty} u_{n_j - d} = \lambda$ for all $d \geq 0$.

At this point it is tempting to say that the limit of the original u_n sequence is λ. However, this does not follow from the fact that $\lim_{j\to\infty} u_{n_j-d} = \lambda$ for all $d \geqslant 0$. To see this consider the following example. Let $w_n = 1$ if $k^2 - k < n \leqslant k^2$ for some integer k, and $w_n = 0$ otherwise. The subsequence w_{n^2} converges to 1. In fact w_{n^2-d} converges to 1 for all $d \geqslant 0$. However w_{n^2+1} converges to zero so $\lim_{n\to\infty} w_n$ does not exist. In view of this example we see that more must be done before we can conclude that $\lim_{n\to\infty} u_n$ exists.

Let $r_n = a_{n+1} + a_{n+2} + \dots$. It follows that

$$\sum_{k=0}^{\infty} r_k = \sum_{k=0}^{\infty} \sum_{j=k+1}^{\infty} a_j = \sum_{j=1}^{\infty} \sum_{k=0}^{j-1} a_j = \sum_{j=1}^{\infty} j a_j$$

so the hypothesis involving $\sum_{j=1}^{\infty} j a_j$ will be given below in terms of the r_k's. (The interchange of the series is justified since all the terms are non-negative.) By construction we have $a_n = r_{n-1} - r_n$ for $n \geqslant 1$ so from $u_n - \sum_{k=0}^{n} a_k u_{n-k} = b_n$ we get

$$u_n - a_0 u_n - \sum_{k=1}^{n} (r_{k-1} - r_k) u_{n-k} = b_n$$

and from this it follows that

$$r_0 u_n + r_1 u_{n-1} + \dots + r_n u_0 = r_0 u_{n-1} + r_1 u_{n-2} + \dots + r_{n-1} u_0 + b_n.$$

Set $A_n = r_0 u_n + \dots + r_n u_0$ and the above equation can be rewritten as $A_n = A_{n-1} + b_n$ where $A_0 = r_0 u_0 = (1 - a_0) u_0 = b_0$. Solving this difference equation, we get $A_n = \sum_{i=0}^{n} b_i$. For any fixed N, less than n_j, we have that

$$r_0 u_{n_j} + r_1 u_{n_j-1} + \dots + r_N u_{n_j-N} \leqslant A_{n_j} = \sum_{n=0}^{n_j} b_n.$$

Letting $j \to \infty$ we get $(r_0 + r_1 + \dots + r_N)\lambda \leqslant \sum_{n=0}^{\infty} b_n$, so $\lambda \leqslant \sum_{n=0}^{\infty} b_n / \sum_{k=0}^{N} r_k$. Hence, if $\sum_{k=0}^{\infty} r_k = \infty$ then λ must be zero.

If $\sum_{k=0}^{\infty} r_k < \infty$, then let $\nu = \underline{\lim}_{n\to\infty} u_n$. As above, we select a subsequence n_k such that $\lim_{k\to\infty} u_{n_k} = \nu$ and show that $\lim_{k\to\infty} u_{n_k-d} = \nu$ for all $d \geqslant 0$. Define $g(N) = \sum_{n=N+1}^{\infty} r_n$ so $\lim_{N\to\infty} g(N) = 0$. Now

$$\sum_{n=0}^{n_k} b_n \leqslant r_0 u_{n_k} + r_1 u_{n_k-1} + \dots + r_N u_{n_k-N} + Mg(N)$$

where M is the supremum of the u_n's. Letting $k \to \infty$ we get

$$\sum_{n=0}^{\infty} b_n \leqslant \sum_{n=0}^{N} r_n \nu + Mg(N).$$

Now let $N \to \infty$ and get $\sum_{n=0}^{\infty} b_n \leqslant \sum_{n=0}^{\infty} r_n \nu$ so that

$$\nu \geqslant \frac{\sum\limits_{n=0}^{\infty} b_n}{\sum\limits_{n=0}^{\infty} r_n}.$$

Combining this with the above we get $\nu \geqslant \lambda$ so $\nu = \lambda$. Hence,

$$\lim_{n \to \infty} u_n = \frac{\sum\limits_{n=0}^{\infty} b_n}{\sum\limits_{n=0}^{\infty} r_n}. \qquad \blacktriangle$$

EXERCISES

1. (a) Consider the set of integers $\{4, 9\}$. Show how 1 can be written as a linear combination of 4 and 9. Find the smallest integer, N, such that if $n \geqslant N$, then n can be written as a linear combination of 4 and 9 with *non-negative* integer coefficients.
   (b) Repeat part (a) using the set of integers $\{6, 10, 15\}$.

2. Let

$$a_k = \left(\tfrac{1}{2}\right)^{k+1}, \qquad b_k = \frac{3}{2^{2k+1}}, \qquad u_k = \frac{(2^k + 1)}{2^k} \quad \text{for } k = 0, 1, 2, \dots.$$

   Show that the hypotheses of the Renewal Theorem (Theorem III.1.1) hold and verify that the conclusion of the theorem holds.

3. Discuss what changes would be needed in the proof of the Renewal Theorem if the assumptions $a_2 > 0$ and $a_3 > 0$ were changed to G.C.D. $\{n : a_n > 0\} = 1$.

SECTION 2   CONSEQUENCES OF THE RENEWAL THEOREM

We now consider what is probably the most important question in the theory of Markov chains. Namely, how will the chain be distributed among the states of S after a "long time"? In particular, will the distribution be asymptotically independent of n? Recall that for fixed n^*, the stochastic process is a random variable, X_{n^*}. The question above asks whether or not the distribution of X_n converges to a limiting distribution as $n \to \infty$. This limiting distribution, when it exists, is called the long run distribution or invariant probability distribution or stationary probability distribution of $\{X_n\}$.

Example III.2.1. Let $\{X_n\}$ be a Markov chain with state space, $S = \{1, 2\}$ and transition matrix, $P = \begin{pmatrix} \frac{1}{2} & \frac{1}{2} \\ \frac{1}{4} & \frac{3}{4} \end{pmatrix}$. Consider the following powers of P:

$$P^2 = \begin{pmatrix} 6/16 & 10/16 \\ 5/16 & 11/16 \end{pmatrix}, \qquad P^4 = \begin{pmatrix} 86/256 & 170/256 \\ 85/256 & 171/256 \end{pmatrix}.$$

As higher powers of P are taken we see that $\lim_{n \to \infty} P^n = \begin{pmatrix} \frac{1}{3} & \frac{2}{3} \\ \frac{1}{3} & \frac{2}{3} \end{pmatrix}$. Hence in this case there is a limiting distribution in the sense that

$$\lim_{n \to \infty} p_{11}^{(n)} = \lim_{n \to \infty} p_{21}^{(n)} = \tfrac{1}{3} \quad \text{and} \quad \lim_{n \to \infty} p_{12}^{(n)} = \lim_{n \to \infty} p_{22}^{(n)} = \tfrac{2}{3}.$$

This implies that no matter where the chain starts, $P[X_n = 1] \to \tfrac{1}{3}$ and $P[X_n = 2] \to \tfrac{2}{3}$ as $n \to \infty$.

The key result that gives this long run distribution in certain cases is logically a simple corollary of the renewal theorem. However, since the result is so basic to Markov chains, it will be presented here as a theorem.

Theorem III.2.1.

Let $P = (p_{ij})$ be the transition matrix for an irreducible, persistent, aperiodic Markov chain. Then for all $i \in S$

(i) $\qquad \lim_{n \to \infty} p_{ii}^{(n)} = \dfrac{1}{\sum_{k=0}^{\infty} k f_{ii}^{(k)}} \qquad \text{if } \sum_{k=0}^{\infty} k f_{ii}^{(k)} < \infty$

$$= 0 \text{ otherwise.}$$

(ii) $\qquad \lim_{n \to \infty} p_{ji}^{(n)} = \dfrac{1}{\sum_{k=0}^{\infty} k f_{ii}^{(k)}} \qquad \text{for all } j \in S \text{ if } \sum_{k=0}^{\infty} k f_{ii}^{(k)} < \infty$

$$= 0 \text{ otherwise.}$$

Proof. Let i be an arbitrary but fixed state in S. In using the renewal theorem to prove this theorem we simply must identify the sequences $\{a_n\}$, $\{b_n\}$, and $\{u_n\}$ in terms of sequences arising from the Markov chain and then check the hypotheses of the renewal theorem. The sequences are chosen as follows: $a_n = \{f_{ii}^{(n)}\}$, $u_n = \{p_{ii}^{(n)}\}$, $b_n = 0$ if $n \geqslant 1$ and $b_0 = 1$. By hypothesis all the states are aperiodic and persistent so in particular state i is. Since state i is aperiodic, we have G.C.D. $\{n : f_{ii}^{(n)} > 0\} = 1$. Since u_n is a probability for each n, we have that the $\{u_n\}$ sequence is bounded. By construction $\sum_{n=0}^{\infty} b_n = 1 < \infty$. Since state i is persistent, we have $\sum_{n=0}^{\infty} f_{ii}^{(n)} = 1$. The renewal equation in this case says

$$p_{ii}^{(n)} - \sum_{k=0}^{n} f_{ii}^{(k)} p_{ii}^{(n-k)} = 0$$

if $n \geqslant 1$ and $p_{ii}^{(0)} = 1$. These equations have been established previously. [See (2.1) in Chapter II.] Hence in applying the renewal theorem we get

$$\lim_{n \to \infty} p_{ii}^{(n)} = \frac{1}{\displaystyle\sum_{k=0}^{\infty} k f_{ii}^{(k)}} \quad \text{when} \quad \sum_{k=0}^{\infty} k f_{ii}^{(k)} < \infty$$

and

$$\lim_{n \to \infty} p_{ii}^{(n)} = 0 \quad \text{when} \quad \sum_{k=0}^{\infty} k f_{ii}^{(k)} = \infty.$$

Note that $\sum_{k=0}^{\infty} k f_{ii}^{(k)}$ equals μ_i, the expected return time to state i, so the limit of $p_{ii}^{(n)}$ is reciprocally related to this expected return time. (Now the designations of positive persistent for states with $\sum_{k=0}^{\infty} k f_{ii}^{(k)} < \infty$ and null persistent for states with $\sum_{k=0}^{\infty} k f_{ii}^{(k)} = \infty$ make some sense. State i is positive persistent if $\lim_{n \to \infty} p_{ii}^{(n)}$ is positive and it is null persistent if $\lim_{n \to \infty} p_{ii}^{(n)}$ is zero.)

To prove part (ii) of the theorem let $\lim_{n \to \infty} p_{ii}^{(n)} = \lambda$ and note that $p_{ji}^{(n)} = \sum_{k=0}^{n} f_{ji}^{(k)} p_{ii}^{(n-k)}$. Since the chain is irreducible and persistent, we have $\sum_{k=0}^{\infty} f_{ji}^{(k)} = 1$. Using Theorem I.4.4 we get

$$\lim_{n \to \infty} p_{ji}^{(n)} = \lim_{n \to \infty} \sum_{k=0}^{n} f_{ji}^{(k)} p_{ii}^{(n-k)} = 1 \cdot \lambda. \quad \blacktriangle$$

We now have established the fact that for irreducible, persistent, aperiodic Markov chains the $\lim_{n \to \infty} p_{ji}^{(n)}$ exists and is independent of j. In fact the value of the limit is given by (ii) of Theorem III.2.1. That is, in the null persistent case the limit is zero and in the positive persistent case the limit is $1 / \sum_{k=0}^{\infty} k f_{ii}^{(k)}$.

The following example shows that for a positive persistent chain, it is not always easy to calculate $\sum_{k=0}^{\infty} k f_{ii}^{(k)}$.

Example III.2.2. Let $\{X_n\}$ be a Markov chain with state space $S = \{1,2,3,4\}$ and transition matrix

$$P = \begin{bmatrix} \frac{1}{2} & \frac{1}{6} & \frac{1}{6} & \frac{1}{6} \\ 0 & 0 & 1 & 0 \\ 0 & 0 & 0 & 1 \\ 1 & 0 & 0 & 0 \end{bmatrix}.$$

By considering state 1, it is easy to see that this chain is persistent, irreducible, and aperiodic. (Why?) However, it is not so easy to calculate $f_{22}^{(k)}$ and then $\sum_{k=0}^{\infty} k f_{22}^{(k)}$. The reader should try to do this and in this attempt, convince himself that it is messy. A second approach to finding the limit would be to simply take powers of P as was done in Example III.2.1. That is, if it is known that $\lim_{n \to \infty} p_{ij}^{(n)} = \pi_j$ exists, then by taking powers of P the (i,j)th entry of P^n for large n is an approximation of π_j. The main disadvantage with this approach is that generally $p_{ij}^{(n)}$ will never equal π_j but only get close to it. A second disadvantage of this approach is that if P is a large matrix, taking powers is a nontrivial task. The reader might try this method on Example III.2.2.

In view of the above example we look for another way to evaluate $\lim_{n \to \infty} p_{ii}^{(n)}$ in the irreducible aperiodic positive persistent case. The following theorem provides a method that in many cases is very simple to use.

Theorem III.2.2.

Let $\{X_n\}$ be an irreducible, aperiodic, positive persistent Markov chain with transition probabilities (p_{ij}). Since $\lim_{n \to \infty} p_{ij}^{(n)}$ is known to exist independently of i in this case, define $\pi_j = \lim_{n \to \infty} p_{ij}^{(n)}$. The π_j's satisfy the following conditions:

$$\pi_j > 0, \qquad \sum_{j \in S} \pi_j = 1, \quad \text{and} \quad \pi_j = \sum_{i \in S} \pi_i p_{ij}. \tag{2.1}$$

Conversely the π_j's are uniquely determined by the restrictions, $\pi_j \geq 0$ and $\pi_j = \sum_{i \in S} \pi_i p_{ij}$, for all $j \in S$ and $\sum_{j \in S} \pi_j = 1$.

Proof. We first prove that the π_j's satisfy (2.1). The positivity of the π_j's follows directly from the fact that the chain is positive persistent.

By the Chapman–Kolmogorov equation we have $p_{ij}^{(n+1)} = \sum_{k \in S} p_{ik}^{(n)} p_{kj}$. If the state space is finite, (2.1) is easy to prove (Exercise 8). If it is true that for all $j \in S$, $\Sigma_{k \in S} p_{kj} < \infty$, then Lebesgue's dominated convergence theorem can be used to take the limit as $n \to \infty$ of both sides of the Chapman–Kolmogorov equation to get $\pi_j = \Sigma_{k \in S} \pi_k p_{kj}$. However, in case neither of these conditions holds, some other approach is required. For such a case, take the state space, S, to be $\{1, 2, 3, \ldots\}$. Using this state space we know that $\Sigma_{k=1}^{\infty} p_{ik}^{(n)} = 1$ by the properties of a stochastic matrix. Hence for all integers M, $\Sigma_{k=1}^{M} p_{ik}^{(n)} \leq 1$. For this finite sum, the limit as $n \to \infty$ can be brought under the summation sign to yield $\Sigma_{k=1}^{M} \pi_k \leq 1$. Since this holds for all M, it follows that $\Sigma_{k=1}^{\infty} \pi_k \leq 1$.

Now

$$p_{ij}^{(n+1)} = \sum_{k=1}^{\infty} p_{ik}^{(n)} p_{kj} \geq \sum_{k=1}^{M} p_{ik}^{(n)} p_{kj}$$

since all the terms are non-negative. Hence taking limits as $n \to \infty$ it follows that $\pi_j \geq \Sigma_{k=1}^{M} \pi_k p_{kj}$ for all M, so that $\pi_j \geq \Sigma_{k=1}^{\infty} \pi_k p_{kj}$. In order to show equality for each j, we sum both sides on j:

$$\sum_{j=1}^{\infty} \pi_j \geq \sum_{j=1}^{\infty} \sum_{k=1}^{\infty} \pi_k p_{kj} = \sum_{k=1}^{\infty} \sum_{j=1}^{\infty} \pi_k p_{kj} \qquad (2.2)$$

by Fubini's theorem since all terms are non-negative.

Since

$$\sum_{k=1}^{\infty} \pi_k \sum_{j=1}^{\infty} p_{kj} = \sum_{k=1}^{\infty} \pi_k,$$

we have

$$1 \geq \sum_{j=1}^{\infty} \pi_j \geq \sum_{j=1}^{\infty} \sum_{k=1}^{\infty} \pi_k p_{kj} = \sum_{k=1}^{\infty} \pi_k$$

so

$$\sum_{j=1}^{\infty} \pi_j = \sum_{j=1}^{\infty} \sum_{k=1}^{\infty} \pi_k p_{kj}. \qquad (2.3)$$

But, since for each j, $\pi_j \geq \Sigma_{k=1}^{\infty} \pi_k p_{kj}$, Equation 2.3 implies equality for each j. That is, if for any j, $\pi_j > \Sigma_{k=1}^{\infty} \pi_k p_{kj}$ then the sums over j will be unequal since for no j does the inequality reverse itself and the series are convergent. This concludes the proof of the fact that $\pi_j = \Sigma_{k=1}^{\infty} \pi_k p_{kj}$ for all $j \in S$. It remains to be shown that $\Sigma_{j=1}^{\infty} \pi_j = 1$.

We have that

$$\pi_k = \sum_{j=1}^{\infty} \pi_j p_{jk} = \sum_{j=1}^{\infty} \left(\sum_{i=1}^{\infty} \pi_i p_{ij} \right) p_{jk}$$

$$= \sum_{i=1}^{\infty} \pi_i \sum_{j=1}^{\infty} p_{ij} p_{jk} = \sum_{i=1}^{\infty} \pi_i p_{ik}^{(2)}$$

where Fubini's theorem was used to interchange the order of summation.

Using induction it follows that $\pi_k = \sum_{i=1}^{\infty} \pi_i p_{ik}^{(n)}$ for $n = 1, 2, 3, \ldots$. We showed earlier that $\sum_{i=1}^{\infty} \pi_i \leqslant 1$ and $p_{ik}^{(n)} \leqslant 1$ so Lebesgue's dominated convergence theorem can be used to take the limit of both sides of the above equation as $n \to \infty$ to get $\pi_k = \sum_{i=1}^{\infty} \pi_i \pi_k$. Since $\pi_k > 0$, it follows that $\sum_{i=1}^{\infty} \pi_i = 1$.

Conversely assume that for all k, $x_k = \sum_{i=1}^{\infty} x_i p_{ik}$, where $\sum_{i=1}^{\infty} x_i = 1$, and $x_i \geqslant 0$ for all i. It follows as above that for all k, $x_k = \sum_{i=1}^{\infty} x_i p_{ik}^{(n)}$ for $n = 1, 2, 3, \ldots$ so taking the limit as $n \to \infty$ we get for all k, $x_k = \sum_{i=1}^{\infty} x_i \pi_k$ by Lebesgue's dominated convergence theorem. (Note that we used $|x_i p_{ik}^{(n)}| \leqslant x_i$ and $\sum_{i=1}^{\infty} x_i < \infty$.) In fact, $\sum_{i=1}^{\infty} x_i = 1$ so for all k, $x_k = \pi_k$.   ▲

This theorem will be used to determine the $\lim_{n \to \infty} p_{ij}^{(n)} = \pi_j$ for all finite Markov chains for which these limits exist. The resulting distribution $\boldsymbol{\pi} = (\pi_1, \pi_2, \ldots)$ is called the long run distribution or the invariant probability distribution or stationary probability distribution for the chain. The terms invariant and stationary come from the fact that by (2.1) we have $\boldsymbol{\pi} P = \boldsymbol{\pi}$. The term probability distribution comes from the fact that $\pi_i \geqslant 0$ and $\sum_{i=1}^{\infty} \pi_i = 1$. (We will see in Chapter IV that the system of equations $\boldsymbol{\pi} P = \boldsymbol{\pi}$ never has a unique solution. The additional constraint $\sum_{i=1}^{\infty} \pi_i = 1$ must be used to determine the invariant *probability* distribution.)

Example III.2.3.   Let

$$P = \begin{bmatrix} \frac{1}{2} & 0 & \frac{1}{2} \\ 1 & 0 & 0 \\ 0 & 1 & 0 \end{bmatrix}$$

be a stochastic matrix for a Markov chain with state space $\{1, 2, 3\}$. Note that the chain is irreducible positive persistent and aperiodic. In order to find the invariant probability distribution, simply solve the system of equations $(\pi_1, \pi_2, \pi_3)P = (\pi_1, \pi_2, \pi_3)$ and $\pi_1 + \pi_2 + \pi_3 = 1$:

$$(\pi_1, \pi_2, \pi_3)\begin{bmatrix} \frac{1}{2} & 0 & \frac{1}{2} \\ 1 & 0 & 0 \\ 0 & 1 & 0 \end{bmatrix} = \left(\frac{\pi_1}{2} + \pi_2, \pi_3, \frac{\pi_1}{2}\right) = (\pi_1, \pi_2, \pi_3).$$

Hence $\pi_1/2 + \pi_2 = \pi_1$, $\pi_3 = \pi_2$, $\pi_1/2 = \pi_3$. Since these three equations are dependent, one must use $\pi_1 + \pi_2 + \pi_3 = 1$ in order to get a unique solution:

$$\pi_1 + \pi_2 + \pi_3 = 2\pi_3 + \pi_3 + \pi_3 = 1$$

so

$$\pi_3 = \tfrac{1}{4}, \qquad \pi_2 = \tfrac{1}{4}, \quad \text{and} \quad \pi_1 = \tfrac{1}{2}.$$

Example III.2.4. Let X_n denote a Markov chain with state space $S = \{0, 1, 2, \dots\}$. Find the invariant probability distribution for this process if the transition probability matrix is given by

$$P = \begin{bmatrix} \frac{1}{2} & \frac{1}{2} & 0 & 0 & \cdots \\ \frac{3}{4} & 0 & \frac{1}{4} & 0 & \cdots \\ 0 & \frac{7}{8} & 0 & \frac{1}{8} & \cdots \\ \vdots & \vdots & \vdots & \vdots & \end{bmatrix}.$$

$(\pi_0, \pi_1, \pi_2, \dots)P = (\pi_0, \pi_1, \dots)$ yields

$$\frac{\pi_0}{2} + \frac{3\pi_1}{4} = \pi_0$$

$$\frac{\pi_0}{2} + \frac{7\pi_2}{8} = \pi_1$$

$$\frac{\pi_1}{4} + \frac{15\pi_3}{16} = \pi_2, \quad \text{and so on.}$$

Hence

$$\pi_1 = \frac{2\pi_0}{3},$$

$$\pi_2 = \frac{8}{7}\left(\pi_1 - \frac{\pi_0}{2}\right) = \frac{4}{21}\pi_0$$

$$\pi_3 = \frac{16}{15}\left(\frac{4}{21} - \frac{1}{6}\right)\pi_0 = \frac{8}{(15)(21)}\pi_0,$$

so

$$\pi_0 + \pi_1 + \pi_2 + \cdots$$

$$= \pi_0 + \frac{2}{3}\pi_0 + \frac{4}{21}\pi_0 + \cdots + \frac{2^{n-1}}{1\cdot 3\cdot 7\cdot 15\cdots(2^n-1)}\pi_0 + \cdots = 1.$$

Therefore

$$\pi_0 = \cfrac{1}{\displaystyle\sum_{k=1}^{\infty} \cfrac{2^{k-1}}{1\cdot 3\cdot 7\cdots(2^k-1)}}$$

and all other π_i's are determined from π_0.

Theorems III.2.1 and III.2.2 were stated and proved for Markov chains that are irreducible, positive persistent, and aperiodic. We next consider possible limit theorems in the case where some or all of these assumptions are dropped. The assumption of positive persistence can be easily dropped since there is no problem in finding $\lim_{n\to\infty} p_{ij}^{(n)}$ when j is a transient state or a null persistent state. By definition, if j is transient, then $\sum_{n=1}^{\infty} p_{jj}^{(n)} < \infty$ and from this it follows that for all i, $\sum_{n=1}^{\infty} p_{ij}^{(n)} < \infty$ (Exercise 10). Hence, $\lim_{n\to\infty} p_{ij}^{(n)} = 0$ for all states i when j is transient. If state j is null persistent, then by Theorem III.2.1 we have that $\lim_{n\to\infty} p_{ij}^{(n)} = 0$ for all states i. (Note that these statements hold without regard to irreducibility and/or periodicity.)

Using the fact that $\lim_{n\to\infty} p_{ij}^{(n)} = 0$ when j is null persistent or transient we can establish the following lemma:

Lemma III.2.1. In a finite irreducible Markov chain all states are positive persistent.

Proof. Let the state space for the Markov chain be $S = \{1, 2, \ldots, k\}$ and assume that state 1 is null persistent. This implies that all of the states in S are null persistent by Corollary II.2.1. Hence $\lim_{n\to\infty} p_{1j}^{(n)} = 0$ for all j. But since P^n is stochastic for all n, we must have $\sum_{j=1}^{k} p_{1j}^{(n)} = 1$ for all n and this is impossible. (That is there are only finitely many terms in the first row of P^n and since each of these terms goes to zero as $n\to\infty$, the sum cannot equal one for all n.) Hence all the states must be positive persistent. ▲

The above proof also shows that not all states of a finite chain can be transient. [However, while there can be no null persistent states in a finite Markov chain, there can be some transient states (Exercise 14).]

The fact that $\lim_{n\to\infty} p_{ij}^{(n)}$ is always zero when j is transient does allow us to extend Theorem III.2.2 to some finite chains that are not irreducible. The extension can be made to the case where there is one irreducible closed set of positive persistent aperiodic states and the remaining states are transient. To see why this is true recall that if P is the transition matrix of a finite Markov chain with exactly one irreducible closed set, C_1, of positive persistent states, then by reordering the states in S, P can be written as

$$P^* = \begin{pmatrix} P_1 & 0 \\ R & Q \end{pmatrix}.$$

(For notational convenience it will be assumed that P itself has this block form.) The matrix P_1 corresponds to transitions among the aperiodic positive persistent set of states, C_1, and Q corresponds to transitions among the transient states. Since P_1 is a finite stochastic matrix for an irreducible aperiodic positive persistent Markov chain, Theorem III.2.2 applies. Hence $\lim_{n\to\infty} p_{ij}^{(n)} = \pi_j$ if i and j belong to C_1. Also $\lim_{n\to\infty} p_{ij}^{(n)} = 0$ if j is a transient state. It will follow that $\lim_{n\to\infty} p_{ij}^{(n)}$ exists independently of i if we can show that $\lim_{n\to\infty} p_{ij}^{(n)} = \pi_j$ when i is transient and j belongs to C_1.

For a finite chain it is impossible to stay forever among the transient states (see Section 3). Hence with probability one the chain eventually visits the persistent state j from the transient state i and so $\sum_{n=1}^{\infty} f_{ij}^{(n)} = 1$. Using the fact that $p_{ij}^{(n)} = \sum_{k=0}^{n} f_{ij}^{(k)} p_{jj}^{(n-k)}$ and applying Theorem I.4.4 we get $\lim_{n\to\infty} p_{ij}^{(n)} = \pi_j$. This concludes the proof of the fact that the long run distribution exists for a finite Markov chain with one irreducible aperiodic closed set. By following the proof of Theorem III.2.2, it is easy to show that the π_j's satisfy the conditions

(i) $\pi_j \geq 0$ for $j = 1, 2, \ldots, n$.

(ii) $\pi_j = \sum_{i=1}^{n} \pi_i p_{ij}$ for $j = 1, 2, \ldots, n$

(iii) $\sum_{j=1}^{n} \pi_j = 1$.

(In order to prove that this last condition is satisfied, there must be at least one nonzero π_j. This follows from the assumption that C_1 is positive persistent.) It also follows that the unique non-negative solution to the system of equations

$$\text{(i)} \quad x_j = \sum_{i=1}^{n} x_i p_{ij} \quad j = 1, 2, \ldots, n$$

$$\text{(ii)} \quad \sum_{j=1}^{n} x_j = 1$$

is the long run distribution for P.

The next example is an application of the above discussion.

Example III.2.5. Consider a Markov chain with state space $\{1, 2, 3, 4\}$ and transition matrix

$$P = \begin{pmatrix} \frac{1}{4} & 0 & \frac{3}{4} & 0 \\ 0 & \frac{1}{2} & \frac{1}{2} & 0 \\ \frac{1}{3} & 0 & \frac{2}{3} & 0 \\ 0 & \frac{1}{4} & \frac{1}{4} & \frac{1}{2} \end{pmatrix}.$$

P can be written in block form as

$$P^* = \begin{pmatrix} \frac{1}{4} & \frac{3}{4} & 0 & 0 \\ \frac{1}{3} & \frac{2}{3} & 0 & 0 \\ 0 & \frac{1}{2} & \frac{1}{2} & 0 \\ 0 & \frac{1}{4} & \frac{1}{4} & \frac{1}{2} \end{pmatrix}$$

where the rows and columns correspond to states, $1, 3, 2, 4$ respectively. In this case

$$P_1 = \begin{pmatrix} \frac{1}{4} & \frac{3}{4} \\ \frac{1}{3} & \frac{2}{3} \end{pmatrix} \quad \text{and} \quad Q = \begin{pmatrix} \frac{1}{2} & 0 \\ \frac{1}{4} & \frac{1}{2} \end{pmatrix}.$$

The invariant distribution for P_1 is $\pi_1 = 4/13$ and $\pi_2 = 9/13$. Hence the invariant distribution for P^* is $(4/13, 9/13, 0, 0)$ and the invariant distribution for P is $(4/13, 0, 9/13, 0)$. Alternatively the long run distribution for P can be determined without first reducing P to block form. To do this simply solve the system of equations $\pi P = \pi$ and $\sum_{i=1}^{4} \pi_i = 1$. In this case it is certainly easier to find the long run distribution without first reducing P to block form. The reason that P was reduced to block form was so that the example would relate to the earlier discussion about matrices in block form. (Note that the terms long run distribution, invariant probability distribution, and stationary probability distribution will be used interchangeably from now on.)

***Remark* III.2.1.** If a Markov chain with one irreducible aperiodic positive persistent class has infinitely many transient states, the long run distribution might not exist since in such cases the chain can stay forever among the transient states (see Section 3).

We now return to the general question of when $\lim_{n\to\infty} p_{ij}^{(n)}$ exists independently of i. At this point we know that if state j is transient or null persistent, this limit is zero without regard to irreducibility and/or aperiodicity. The next two examples show that if j is positive persistent, both irreducibility and aperiodicity are necessary in order for $\lim_{n\to\infty} p_{ij}^{(n)}$ to exist independently of i.

Example III.2.6. Let $\{X_n\}$ be a Markov chain with state space $S = \{1,2,3,4\}$ and transition matrix

$$P = \begin{bmatrix} 1 & 0 & 0 & 0 \\ 0 & 1 & 0 & 0 \\ \frac{1}{3} & \frac{2}{3} & 0 & 0 \\ \frac{1}{4} & \frac{1}{4} & 0 & \frac{1}{2} \end{bmatrix}.$$

It is easy to show that this chain is aperiodic but *not* irreducible. In this case $p_{11}^{(n)} = 1$ for all n, $p_{21}^{(n)} = 0$ for all n, $p_{31}^{(n)} = \frac{1}{3}$ for all n, and $\lim_{n\to\infty} p_{41}^{(n)} = \frac{1}{2}$. Hence the limit of $p_{i1}^{(n)}$ exists but it certainly depends on state i.

Example III.2.7. Let $\{X_n\}$ be a Markov chain with state space, $S = \{1,2\}$ and transition matrix

$$P = \begin{pmatrix} 0 & 1 \\ 1 & 0 \end{pmatrix}.$$

It is easy to see that this chain is irreducible but *not* aperiodic. In this case $p_{11}^{(n)} = 0$ if n is odd and 1 if n is even. Hence $\lim_{n\to\infty} p_{11}^{(n)}$ does not exist. However if attention is restricted to only odd times or only even times, then the corresponding subsequences converge. In particular, $\lim_{n\to\infty} p_{12}^{(2n)} = 0$ and $\lim_{n\to\infty} p_{12}^{(2n+1)} = 1$.

The above property of converging subsequences will be shown to hold for all irreducible Markov chains. In order to show this the following lemma is needed.

Lemma III.2.2. The state space of a periodic irreducible Markov chain of period d can be partitioned into d disjoint classes $D_0, D_1, \ldots, D_{d-1}$ such that from D_j the chain goes, in the next step, to D_{j+1} for $j = 0, 1, \ldots, d-2$. From D_{d-1} the chain returns in the next step to D_0.

Proof.   Let i be any fixed state in S. Define $D_m = \{j \mid p_{ij}^{(nd+m)} > 0$ for some $n\}$. Since the chain is irreducible, we have $\cup_{m=0}^{d-1} D_m = S$. In order to show that the D_m's are disjoint, let $j \in D_{m_1} \cap D_{m_2}$ with $0 \leqslant m_1 < d$, $0 \leqslant m_2 < d$. Since $j \in D_{m_1}$, there exists n_1 such that $p_{ij}^{(n_1 d + m_1)} > 0$. Also since the chain is irreducible there exists k such that $p_{ji}^{(k)} > 0$. Hence $p_{ii}^{(n_1 d + m_1 + k)} \geqslant p_{ij}^{(n_1 d + m_1)} p_{ji}^{(k)} > 0$. Similarly $p_{ii}^{(n_2 d + m_2 + k)} \geqslant p_{ij}^{(n_2 d + m_2)} p_{ji}^{(k)} > 0$. It follows that $m_1 + k$ is a multiple of d and $m_2 + k$ is a multiple of d. Hence $(m_1 + k) - (m_2 + k) = m_1 - m_2$ is a multiple of d but since m_1 and m_2 lie between zero and $d - 1$ we have $m_1 = m_2$. So if D_{m_1} and D_{m_2} have an element in common, they are equal. Hence the D_m's form a partition of S and by construction the chain moves from D_j to D_{j+1}. ▲

Example III.2.8.   Let $\{X_n\}$ be a Markov chain with transition matrix

$$P = \begin{bmatrix} 0 & 0 & \frac{1}{2} & 0 & \frac{1}{2} & 0 \\ \frac{1}{3} & 0 & 0 & \frac{1}{3} & 0 & \frac{1}{3} \\ 0 & 1 & 0 & 0 & 0 & 0 \\ 0 & 0 & 1 & 0 & 0 & 0 \\ 0 & 1 & 0 & 0 & 0 & 0 \\ 0 & 0 & \frac{1}{4} & 0 & \frac{3}{4} & 0 \end{bmatrix}.$$

This chain is irreducible and periodic of period 3. Starting with state 1, the following disjoint classes are determined. $D_0 = \{j: p_{1j}^{(3n)} > 0$ for some $n\} = \{1, 4, 6\}$, $D_1 = \{j: p_{1j}^{(3n+1)} > 0$ for some $n\} = \{3, 5\}$, and $D_2 = \{j: p_{1j}^{(3n+2)} > 0$ for some $n\} = \{2\}$. The chain moves deterministically from class D_0 to D_1 to D_2 to D_0 and so on. The actual state visited within each class is still random.

If P is a stochastic matrix of period d, then $R = P^d$ maps any of the sets D_m given by Lemma III.2.2 into itself. In fact, using $R = (r_{ij})$ as a new transition probability matrix each D_m can be viewed as a state space for an irreducible *aperiodic* Markov chain. The following theorem gives the value of $\lim_{n \to \infty} r_{ij}^{(n)}$.

Theorem III.2.3.

Let P be the transition matrix for an irreducible periodic positive persistent Markov chain with period d. Let $R = P^d$. Then

$$\lim_{n \to \infty} r_{ij}^{(n)} = \frac{d}{\mu_j} \quad \text{if } i \text{ and } j \text{ belong to the same class, } D_m,$$

and

$$\lim_{n\to\infty} r_{ij}^{(n)} = 0 \quad \text{otherwise,}$$

where μ_j denotes the expected recurrence time to state j using the transition matrix P.

Proof. Since R operates like a transition matrix for an irreducible, aperiodic, positive persistent Markov chain, we know that $\lim_{n\to\infty} r_{ii}^{(n)}$ is the reciprocal of the mean recurrence time, ν_i, relative to R. That is $\nu_i = \sum_{k=1}^{\infty} k h_{ii}^{(k)}$ where $h_{ii}^{(k)} = P[\text{first return to } i \text{ at step } k \text{ using the transition matrix } R]$. In general $\nu_i \neq \mu_i$ since the concept of mean recurrence time does depend upon whether the transition matrix P or R is used. In particular if the transition matrix P is used, d time units pass in going from X_0 to X_d. If R is used, only one time unit passes between X_0 and X_d. These different methods of marking time clearly yield different values for the mean recurrence time to state i.

Now

$$\sum_{k=1}^{\infty} k h_{ii}^{(k)} = \sum_{k=1}^{\infty} k f_{ii}^{(dk)} = \frac{1}{d} \sum_{k=1}^{\infty} dk f_{ii}^{(dk)} = \frac{1}{d} \sum_{n=1}^{\infty} n f_{ii}^{(n)}$$

since $f_{ii}^{(n)} = 0$ unless $n = dk$. This last expression is equal to (μ_i/d), so $\lim_{n\to\infty} r_{ii}^{(n)} = d/\mu_i$. If i and j belong to the same D_m, then $\lim_{n\to\infty} r_{ij}^{(n)} = 1/\nu_j = d/\mu_j$. Obviously if i and j belong to different classes, then $\lim_{n\to\infty} r_{ij}^{(n)} = 0$.
▲

Remark III.2.2. If i and j belong to different classes, then $\lim_{k\to\infty} p_{ij}^{(kd+m)} = d/\mu_j$ for m properly chosen (Exercise 11).

For the general case of irreducible periodic chains, we have established the convergence of $p_{ij}^{(n)}$ along a properly chosen subsequence. The following theorem shows that Cesaro averages also converge.

Theorem III.2.4.

Let P be the transition probability matrix for an irreducible, positive persistent Markov chain. Then $(1/n)\sum_{k=1}^{n} p_{ij}^{(k)}$ converges as $n\to\infty$ and the limit is $1/\mu_j$ where μ_j is the expected recurrence time to state j. The vector $\psi = (1/\mu_1, 1/\mu_2, \ldots)$ is an invariant probability vector for P.

Proof. Assume the chain has period d.

For notational convenience we first assume i and j belong to the same class. Let $[n/d]$ denote the greatest integer contained in n/d.

$$\frac{1}{n}\sum_{k=1}^{n} p_{ij}^{(k)} = \frac{1}{n}\sum_{k=1}^{[n/d]} p_{ij}^{(kd)}$$

$$= \frac{1}{d}\left(\frac{d}{n}\right)\sum_{k=1}^{[n/d]} p_{ij}^{(kd)} \to \frac{1}{d}\cdot\frac{d}{\mu_j} = \frac{1}{\mu_j}.$$

To see this note that if $p_{ij}^{(kd)}\to d/\mu_j$, then (Exercise 9)

$$\frac{1}{\left[\dfrac{n}{d}\right]}\sum_{k=1}^{[n/d]} p_{ij}^{(kd)} \to \frac{d}{\mu_j} \quad \text{and} \quad \left[\frac{n}{d}\right]\left(\frac{d}{n}\right)\to 1 \text{ as } n\to\infty. \qquad (2.4)$$

If i and j belong to different classes, consider

$$\frac{1}{n}\sum_{k=1}^{[n/d]} p_{ij}^{(kd+m)}$$

where m is the minimum number of steps required to go from the class containing state i to the class containing state j.

The proof that ψ is an invariant probability vector for P is left as an exercise. ▲

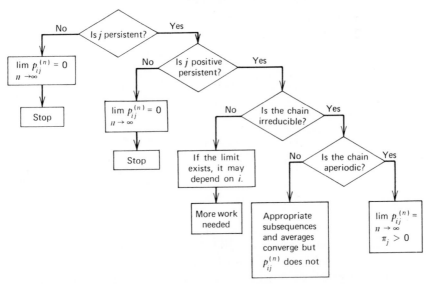

Figure III.2.1. Flow chart for determining $\lim_{n\to\infty} p_{ij}^{(n)}$.

Figure III.2.1 shows the results obtained thus far regarding the limit of $p_{ij}^{(n)}$ as $n \to \infty$ for discrete-time stationary Markov chains.

EXERCISES

1. For the Markov chain defined in Example III.2.2:
   (a) Show that this chain is persistent, irreducible, and aperiodic by considering state 1.
   (b) Calculate P^2, P^4, and P^8. Use these results to estimate π_j, $j = 1, 2, 3, 4$.
   (c) Calculate $\{\pi_j\}_{j=1}^4$ by solving $\pi P = \pi$ and using $\sum_{j=1}^4 \pi_j = 1$.

2. Let $\{X_n\}$ be a Markov chain with state space $S = \{1, 2, 3, 4, 5\}$ and transition matrix

$$P = \begin{bmatrix} 0 & 1 & 0 & 0 & 0 \\ \frac{1}{2} & 0 & 0 & \frac{1}{2} & 0 \\ \frac{1}{3} & 0 & \frac{2}{3} & 0 & 0 \\ 1 & 0 & 0 & 0 & 0 \\ \frac{1}{4} & 0 & \frac{1}{4} & 0 & \frac{1}{2} \end{bmatrix}.$$

   (a) Classify all the states of this chain.
   (b) Find the long run distribution for this chain.
   (c) Find the mean recurrence time to all the persistent states.

3. Let $\{X_n\}$ be a Markov chain with state space $S = \{1, 2, 3, 4, 5, 6, 7, 8\}$ and transition matrix

$$P = \begin{bmatrix} 0 & \frac{1}{2} & 0 & 0 & \frac{1}{2} & 0 & 0 & 0 \\ \frac{1}{3} & 0 & \frac{1}{3} & 0 & 0 & \frac{1}{3} & 0 & 0 \\ \frac{2}{3} & 0 & \frac{1}{6} & 0 & 0 & \frac{1}{6} & 0 & 0 \\ 0 & 0 & 0 & 1 & 0 & 0 & 0 & 0 \\ 0 & \frac{1}{2} & 0 & 0 & 0 & 0 & 0 & \frac{1}{2} \\ 0 & 0 & \frac{1}{2} & 0 & 0 & \frac{1}{2} & 0 & 0 \\ 0 & 0 & 0 & 0 & 0 & 0 & 1 & 0 \\ 0 & 0 & 0 & 0 & \frac{1}{2} & 0 & 0 & \frac{1}{2} \end{bmatrix}.$$

   (a) Classify the states of this chain.
   (b) Find an invariant left probability vector for P. (That is, find a probability vector, ψ, such that $\psi P = \psi$.)
   (c) Does this probability vector represent the long run distribution?
   (d) Note that the rows and columns of P sum to one. A non-

negative matrix with this property is called *doubly stochastic*. All doubly stochastic matrices have one invariant left probability vector in common. Is the vector given as your answer to part b of this problem invariant under all 8×8 doubly stochastic matrices? If not, find the one that is.

*4. Let $\{X_n\}$ be a Markov chain with state space, $S = \{0, 1, 2, \ldots\}$ and transition matrix

$$P = \begin{pmatrix} 0 & 1 & 0 & 0 & 0 & 0 & \cdots \\ \frac{1}{4} & 0 & \frac{3}{4} & 0 & 0 & 0 & \cdots \\ \frac{1}{4} & 0 & 0 & \frac{3}{4} & 0 & 0 & \cdots \\ \frac{1}{4} & 0 & 0 & 0 & \frac{3}{4} & 0 & \cdots \\ \vdots & \vdots & \vdots & \vdots & \vdots & \vdots & \vdots \vdots \vdots \end{pmatrix}$$

(a) Classify the states of this chain.
(b) Find the long run distribution if it exists.

*5. Let $\{X_n\}$ be a Markov chain with state space $S = \{1, 2, 3, \ldots\}$ and transition matrix

$$P = \begin{pmatrix} \frac{1}{2} & \frac{1}{2} & 0 & 0 & 0 & 0 & \cdots \\ \frac{1}{3} & 0 & \frac{2}{3} & 0 & 0 & 0 & \cdots \\ \frac{1}{4} & 0 & 0 & \frac{3}{4} & 0 & 0 & \cdots \\ \frac{1}{5} & 0 & 0 & 0 & \frac{4}{5} & 0 & \cdots \\ \frac{1}{6} & 0 & 0 & 0 & 0 & \frac{5}{6} & \cdots \\ \vdots & \vdots & \vdots & \vdots & \vdots & \vdots & \vdots \vdots \vdots \end{pmatrix}.$$

(a) Classify the states of this chain.
(b) Find the long run distribution if it exists.
(c) Is there an invariant left probability vector for P?

6. Let $\{X_n\}$ be a Markov chain with state space $S = \{1, 2, 3, \ldots\}$ and transition matrix

$$P = \begin{pmatrix} 0 & 1 & 0 & 0 & 0 & 0 & \cdots \\ \frac{1}{2} & 0 & \frac{1}{2} & 0 & 0 & 0 & \cdots \\ 0 & \frac{1}{2} & 0 & \frac{1}{2} & 0 & 0 & \cdots \\ 0 & 0 & \frac{1}{2} & 0 & \frac{1}{2} & 0 & \cdots \\ 0 & 0 & 0 & \frac{1}{2} & 0 & \frac{1}{2} & \cdots \\ \vdots & \vdots & \vdots & \vdots & \vdots & \vdots & \vdots \vdots \vdots \end{pmatrix}$$

(a) Classify the states of this chain.

(b) Find the long run distribution if it exists.

(Note: This chain is often called a simple random walk with reflecting barrier.) Hint: use Exercise III.3.7.

7. Let $\{X_n\}$ be a Markov chain with state space $S = \{1, 2, 3, \dots\}$ and transition matrix

$$P = \begin{pmatrix}
1 & 0 & 0 & 0 & 0 & 0 & 0 & 0 & \cdots \\
0 & \frac{1}{2} & \frac{1}{2} & 0 & 0 & 0 & 0 & 0 & \cdots \\
0 & \frac{1}{2} & 0 & \frac{1}{2} & 0 & 0 & 0 & 0 & \cdots \\
0 & \frac{1}{4} & 0 & 0 & \frac{3}{4} & 0 & 0 & 0 & \cdots \\
0 & \frac{1}{8} & 0 & 0 & 0 & \frac{7}{8} & 0 & 0 & \cdots \\
0 & \frac{1}{16} & 0 & 0 & 0 & 0 & \frac{15}{16} & 0 & \cdots \\
\vdots & \vdots & \vdots & \vdots & \vdots & \vdots & \vdots & \vdots & \vdots
\end{pmatrix}$$

(a) Classify the states of this chain.

(b) Find the long run distribution if it exists.

(c) Is there an invariant left probability vector for P?

8. Show (2.1) of Theorem III.2.2 in the case where S is finite.

9. Show that $\left[\dfrac{n}{d}\right]\left(\dfrac{d}{n}\right) \to 1$ [see (2.4)].

10. Show that if state j is transient, then $\sum_{n=1}^{\infty} p_{ij}^{(n)} < \infty$ for all states i.

11. Show that if i and j are arbitrary states in an irreducible positive persistent discrete-time stationary Markov chain of period d, then $\lim_{k\to\infty} p_{ij}^{(kd+m)} = d/\mu_j$ for some m $(0 \leqslant m < d)$.

12. Show that if $\{X_n\}$ is an irreducible positive persistent Markov chain, then $\{1/\mu_j\}$ is an invariant probability vector for the transition matrix P.

*13. Prove that if one state in an irreducible aperiodic chain is positive persistent, then all the states are positive persistent.

*14. Show that in a finite Markov chain there can be no null persistent states but there can be some transient states.

15. Using the matrix P given in Example I.3.7, find $\lim_{k\to\infty} \alpha_k$, where α_k is as defined in that example.

16. Let $\{X_n\}$ be a Markov chain with state space $S = \{1, 2, 3, \dots\}$ and

transition matrix

$$P = \begin{pmatrix} \frac{1}{2} & \frac{1}{2} & 0 & 0 & 0 & 0 & \cdots \\ \frac{1}{2} & 0 & \frac{1}{2} & 0 & 0 & 0 & \cdots \\ \frac{1}{3} & 0 & 0 & \frac{2}{3} & 0 & 0 & \cdots \\ \frac{1}{3} & 0 & 0 & 0 & \frac{2}{3} & 0 & \cdots \\ \frac{1}{4} & 0 & 0 & 0 & 0 & \frac{3}{4} & \cdots \\ \vdots & \vdots & \vdots & \vdots & \vdots & \vdots & \vdots \vdots \vdots \end{pmatrix}$$

(a) Show that the recurrence time for state 1 has a finite mean.
(b) Show that the recurrence time for state 1 has an infinite variance.
(c) Find a Markov chain for which the recurrence time to state 1 has exactly k moments.

17. Let $\{X_n\}$ be a Markov chain with transition matrix P and state space $S = \{1, 2, 3, \ldots, \}$. Define a transition matrix R on $S \times S$ as

$$r_{(i,j)(k,l)} = p_{ik} p_{jl}.$$

(a) Show that $r_{(i,j)(k,l)}^{(n)} = p_{ik}^{(n)} p_{jl}^{(n)}$.
(b) Show that if P is irreducible and aperiodic, then so is R.
(c) Show that if P is positive persistent, then so is R.

SECTION 3   **SUBSTOCHASTIC MATRICES**

From Figure III.2.1. we see that the final case of interest is the case of a reducible chain. It has already been shown (Section 2) that if the state space is finite and consists of some transient states and one irreducible, closed, aperiodic, positive persistent subclass, then $\lim_{n \to \infty} P_{ij}^{(n)} = \pi_j$ exists for all j independently of i and $\sum_{j=1}^{m} \pi_j = 1$. Since the existence of this long run probability distribution is of central importance, we introduce the following definition to classify chains with this property.

Definition III.3.1. *Let $P = (p_{ii})$ be the transition matrix for a Markov chain. If $\lim_{n \to \infty} P_{ij}^{(n)} = \pi_j$ exists for all j independently of i and if $\sum_{j=1}^{\infty} \pi_j = 1$, then we say the chain is ergodic.*

Examples III.2.7 and III.2.6 show respectively that the following conditions are necessary for ergodicity:

1. All the persistent states are aperiodic.
2. There is at most one irreducible closed subset of persistent states.

Actually since the limiting values are required to form a probability distribution, a finite Markov chain is ergodic if and only if there is exactly one irreducible closed subset of positive persistent states and all of these states are aperiodic (Exercise 4). The assumption that the state space is finite is convenient but not necessary for ergodicity. The convenience comes from the fact that in a finite chain the probability of ultimately leaving the transient states and being absorbed into a closed set of persistent states is one (Exercise 5). The probability of ultimately leaving the transient states may or may not be one for a chain with an infinite state space.

Example III.3.1. Let $\{X_n\}$ be a Markov chain with transition matrix

$$P = \begin{pmatrix} 1 & 0 & 0 & 0 & 0 & 0 & \cdots \\ 1 & 0 & 0 & 0 & 0 & 0 & \cdots \\ \frac{1}{3} & \frac{1}{3} & 0 & \frac{1}{3} & 0 & 0 & \cdots \\ 0 & 0 & 0 & 0 & 1 & 0 & \cdots \\ 0 & 0 & 0 & 0 & 0 & 1 & \cdots \\ \vdots & \vdots & \vdots & \vdots & \vdots & \vdots & \vdots\vdots\vdots \end{pmatrix}.$$

In this case state 1 is absorbing while states $\{2,3,4,5,\ldots\}$ are transient. The chain never enters the irreducible closed set $C = \{1\}$ from the transient states $\{4,5,6,\ldots\}$, so the probability of being absorbed is zero if the chain starts in one of these states. The probability of being absorbed from state 3 is $\frac{2}{3}$ and the probability of being absorbed from state 2 is 1.

In cases where a Markov chain stays among the transient states with positive probability there is no long run probability distribution. In these cases the chain "drifts to infinity" with positive probability, so if there is a long run distribution, it is not a *probability* distribution.

Example III.3.2. Let $\{X_n\}$ be a Markov chain with state space $S = \{1,2,3,\ldots\}$ and transition matrix

$$P = \begin{pmatrix} 0 & 1 & 0 & 0 & 0 & \cdots \\ 0 & 0 & 1 & 0 & 0 & \cdots \\ 0 & 0 & 0 & 1 & 0 & \cdots \\ \vdots & \vdots & \vdots & \vdots & \vdots & \vdots\vdots\vdots \end{pmatrix}.$$

For this chain all states are transient and $\lim_{n \to \infty} p_{ij}^{(n)} = 0$ for all j independently of i. This limiting distribution is clearly not a probability distribution on S.

We now consider the case of a Markov chain with one irreducible closed set of positive persistent aperiodic states and infinitely many transient states. When will such a chain be ergodic? Such a chain will be ergodic if and only if the probability of being absorbed into the irreducible closed set of persistent states is one, so the relevant question becomes: What is the probability of staying forever among the transient states of a Markov chain? If the transition matrix is written in block form, $P = \begin{pmatrix} P_1 & O \\ R & Q \end{pmatrix}$ we see that the transition probabilities among the transient states, T, are given by the matrix Q. Noting that the sum of each row of Q is less than or equal to one leads us to the following definition.

Definition III.3.2. *A matrix Q with elements q_{ij} is called substochastic if $q_{ij} \geq 0$ for all i and j and if $\sum_j q_{ij} \leq 1$ for all i.*

The question raised above about the probability of staying forever among the transient states is answered by looking at the substochastic matrix Q corresponding to transitions among the transient states. In view of this it would seem natural to simply take Q from $P = \begin{pmatrix} P_1 & O \\ R & Q \end{pmatrix}$ and consider the above question. There is a subtle problem with this approach. The matrix Q, by itself, does not have the usual probabilistic interpretations. That is, since the sum of some of the rows is less than one, the entries in Q, by themselves, should not be viewed as probabilities. However, as a submatrix of P, the entries in Q do have probabilistic interpretations. As a compromise between using all of P and using Q alone, we embed Q into a stochastic matrix with one absorbing state. This is done since, for the question of staying forever among the transient states, the properties of P_1 are unimportant so we simply take $P_1 = 1$. In this case R is a column vector with the entry $\xi_i = 1 - \sum_{j \in T} q_{ij}$. The stochastic matrix in question now has the form

$$P^* = \begin{bmatrix} 1 & 0 & 0 & 0 & \cdots \\ \xi_1 & & & & \\ \xi_2 & & Q & & \\ \vdots & & & & \end{bmatrix}. \tag{3.1}$$

$(P^*)^2$ has the form

$$(P^*)^2 = \begin{pmatrix} 1 & 0 & 0 & 0 & \cdots \\ \xi_1^{(2)} & & & & \\ \xi_2^{(2)} & & Q^2 & & \\ \vdots & & & & \end{pmatrix}$$

and similarly

$$(P^*)^n = \begin{pmatrix} 1 & 0 & 0 & 0 & \cdots \\ \xi_1^{(n)} & & & & \\ \xi_2^{(n)} & & Q^n & & \\ \vdots & & & & \end{pmatrix}.$$

As was shown in Chapter I, if $q_{ij}^{(n)}$ denotes the probability of going from i to j in n steps, then $q_{ij}^{(n)}$ is the (ij)th entry of Q^n. It also follows that $q_{ij}^{(n+1)} = \sum_k q_{ik} q_{kj}^{(n)}$ for all transient states i and j. The sum over k can be taken either over all states or only over the transient states since $q_{kj}^{(n)} = 0$ when k and j correspond to persistent and transient states respectively. For notational purposes, we again will let T denote the transient states of P^*. The submatrix Q defines transitions among these states.

Definition III.3.3.   *Let Q be the substochastic matrix for transitions among the transient states, T, of a Markov chain. Define $\sigma_i^{(n)}$ to be the probability of staying among the transient states for n steps starting at state i. (Note that $\sigma_i^{(n)} = \sum_{j \in T} q_{ij}^{(n)}$.)*

We now return to the original question of whether or not a Markov chain with one irreducible closed set of positive persistent aperiodic states is ergodic. If there are infinitely many states in the subset of persistent states, the transition matrix cannot be written in the convenient form $\begin{pmatrix} P_1 & 0 \\ R & Q \end{pmatrix}$. However, using the above convention we consider the transition matrix

$$P^* = \begin{pmatrix} 1 & 0 & 0 & \cdots \\ \xi_1 & & & \\ \xi_2 & & Q & \\ \vdots & & & \end{pmatrix}.$$

P^* is ergodic if and only if the probability of staying forever among the transient states, T, is zero. Hence it must be shown that $\lim_{n \to \infty} \sigma_i^{(n)} = 0$ for all states $i \in T$.

Before finding the limit of $\{\sigma_i^{(n)}\}$ we consider the question of whether or not this limit exists. We know that $q_{ij}^{(n+1)} = \sum_{k \in T} q_{ik} q_{kj}^{(n)}$. Sum both sides of this equation over $j \in T$ and use Fubini's theorem to interchange the order of summation to get $\sigma_i^{(n+1)} = \sum_{k \in T} q_{ik} \sigma_k^{(n)}$. Define $\sigma_i^{(0)} = 1$ for all i. From this it follows that $\sigma_i^{(1)} = \sum_{k \in T} q_{ik} \leqslant 1 = \sigma_i^{(0)}$. Using induction it can be shown that for all states $i \in T$, $\sigma_i^{(n+1)} \leqslant \sigma_i^{(n)}$ (Exercise 3). Hence since $0 \leqslant \sigma_i^{(n)} \leqslant 1$ and the sequence is monotone, the limit of $\sigma_i^{(n)}$ exists as $n \to \infty$.

Let $\sigma_i = \lim_{n \to \infty} \sigma_i^{(n)}$. Since $|q_{ik} \sigma_k^{(n)}| \leqslant q_{ik}$ and $\sum_{k \in T} q_{ik} < \infty$, Lebesgue's dominated convergence theorem can be applied to the equation $\sigma_i^{(n+1)} = \sum_{k \in T} q_{ik} \sigma_k^{(n)}$ to yield $\sigma_i = \sum_{k \in T} q_{ik} \sigma_k$. If this system of equations had a unique solution, we would have a convenient way of finding $\{\sigma_i\}_{i \in T}$. Unfortunately there are, in general, many solutions to this equation, so additional information must be used in order to find the solution corresponding to the $\{\sigma_i\}$.

Let $\{x_i\}$ be any solution of the system $x_i = \sum_{k \in T} q_{ik} x_k$ with $0 \leqslant x_i \leqslant 1$. Then for all states $i \in T$, $0 \leqslant x_i \leqslant \sigma_i^{(0)} = 1$ so $x_i = \sum_{k \in T} q_{ik} x_k \leqslant \sum_{k \in T} q_{ik} \sigma_k^{(0)} = \sigma_i^{(1)}$. Continuing in this manner it can be shown that $x_i \leqslant \sigma_i^{(n)}$ for all $i \in T$ and for all n so $x_i \leqslant \sigma_i$ for all $i \in T$. This implies that among all solutions of $x_i = \sum_{k \in T} q_{ik} x_k$ with $0 \leqslant x_i \leqslant 1$, the $\{\sigma_i\}$ solution is the *maximal* solution.

The above discussion is summarized by the following theorem.

Theorem III.3.1.

The probability of staying forever among the transient states, T, of a Markov chain is given by the maximal solution of $x_i = \sum_{k \in T} q_{ik} x_k$ where $0 \leqslant x_i \leqslant 1$ for all i.

Example III.3.3. Consider the transition matrix of Example III.3.1. In this case the system of equations is

$$x_2 = 0, \qquad x_3 = \tfrac{1}{3} x_2 + \tfrac{1}{3} x_4, \qquad x_4 = x_5, \qquad x_5 = x_6, \quad \text{etc.}$$

Hence the maximal solution of this with $0 \leqslant x_i \leqslant 1$ is $x_2 = 0$, $x_3 = \tfrac{1}{3}$, and $x_i = 1$ for $i \geqslant 4$. Note that these values coincide with the answers given in Example III.3.1.

Example III.3.4. Let $\{X_n\}$ be a Markov chain with state space $S = \{1, 2, 3, \ldots\}$ and transition matrix

$$P = \begin{pmatrix} 1 & 0 & 0 & 0 & 0 & \cdots \\ \frac{1}{2} & 0 & \frac{1}{2} & 0 & 0 & \cdots \\ \frac{1}{3} & 0 & 0 & \frac{2}{3} & 0 & \cdots \\ \frac{1}{4} & 0 & 0 & 0 & \frac{3}{4} & \cdots \\ \vdots & \vdots & \vdots & \vdots & \vdots & \vdots \end{pmatrix}.$$

Consider solutions of $x_i = \sum_{j=2}^{\infty} q_{ij} x_j$ with $0 \leqslant x_i \leqslant 1$ where $i = 2, 3, \ldots$. In this case we have $x_2 = \frac{1}{2} x_3$, $x_3 = \frac{2}{3} x_4$, $x_4 = \frac{3}{4} x_5$, and so on. This implies $x_3 = 2x_2$, $x_4 = \frac{3}{2} x_3$, $x_5 = \frac{4}{3} x_4$, so $x_5 = \frac{4}{3} \cdot \frac{3}{2} \cdot 2x_2 = 4x_2$ and in general $x_n = (n - 1)x_2$. Now if x_2 is positive, it is impossible to keep all the x_n's bounded by one. Hence $x_2 = 0$ and so $x_n = 0$ for $n = 2, 3, 4, \ldots$. When the zero solution is the only solution it is the maximal solution so for this example the chain is absorbed with probability one.

In summary, if a Markov chain contains one irreducible closed subset of positive persistent aperiodic states and the chain leaves the transient states with probability one, then the chain is ergodic. That is, $\lim_{n \to \infty} p_{ij}^{(n)} = \pi_j$ exists for all j independently of i. The limit, π_j, will be zero if and only if j is transient. This concludes the analysis as given in Fig. III.2.1 as far as ergodic Markov chains are concerned.

EXERCISES

1. Let $\{X_n\}$ be a Markov chain with transition matrix

$$P = \begin{pmatrix} 1 & 0 & 0 & 0 \\ \frac{1}{3} & \frac{1}{2} & 0 & \frac{1}{6} \\ \frac{1}{2} & \frac{1}{4} & \frac{1}{4} & 0 \\ 0 & \frac{1}{2} & 0 & \frac{1}{2} \end{pmatrix}.$$

   (a) What system of equations must be solved in order to find the probability of absorption from the transient states?
   (b) Find all non-negative solutions to this system.
   (c) What is the probability of staying forever among the transient states?

2.   Repeat problem 1 for the transition matrix

$$P = \begin{pmatrix} 1 & 0 & 0 & 0 & 0 & 0 & 0 & \cdots \\ \frac{1}{2} & 0 & \frac{1}{2} & 0 & 0 & 0 & 0 & \cdots \\ \frac{1}{4} & 0 & 0 & \frac{3}{4} & 0 & 0 & 0 & \cdots \\ \frac{1}{8} & 0 & 0 & 0 & \frac{7}{8} & 0 & 0 & \cdots \\ \frac{1}{16} & 0 & 0 & 0 & 0 & \frac{15}{16} & 0 & \cdots \\ \vdots & \vdots & \vdots & \vdots & \vdots & \vdots & \vdots & \vdots\vdots\vdots \end{pmatrix}.$$

3.   Show that $\sigma_i^{(n)}$ is monotone nonincreasing as $n \to \infty$.

*4.   Show that a finite Markov chain is ergodic if and only if there is exactly one irreducible closed subset of positive persistent, aperiodic states.

5.   Show that if the state space, S, is finite, the probability of staying forever among the transient states is zero.

6.   Show that if there is one irreducible, closed, aperiodic subset, C, of S and if the probability of being absorbed into C is one, then $\lim_{n \to \infty} p_{ij}^{(n)}$ exists independently of i.

7.   Let $\{X_n\}$ be an irreducible Markov chain with state space, S $= \{1, 2, 3, \ldots\}$. Show that state 1 is persistent if and only if the system $x_i = \sum_{j=2}^{\infty} p_{ij} x_j$ for $i = 2, 3, \ldots$ has no solution with $0 \leqslant x_i \leqslant 1$ except $x_i = 0$ for $i = 2, 3, 4, \ldots$.

SECTION 4   **ABSORPTION TIMES AND ABSORPTION PROBABILITIES**

The general question of moving from the transient states into a closed subset of persistent states is also of interest in nonergodic chains. For example, consider a Markov chain with two irreducible closed sets of persistent states and some transient states. For convenience assume that the state space is finite and hence that the transition matrix can be written in the form

$$P = \begin{bmatrix} P_1 & 0 & 0 \\ 0 & P_2 & 0 \\ R_1 & R_2 & Q \end{bmatrix}$$

where P_1, P_2, R_1, R_2, and Q have the usual interpretations (see Section II.2). Since the state space is finite, the chain leaves the transient states with probability one. The question we now ask is: Where does it go? The chain may be absorbed into either C_1 or C_2 so it is of interest to determine the probability of being absorbed into each closed set. Once it is known that a chain will be absorbed from the transient states into a set of persistent states, we next ask how long it will take. That is, what is the expected waiting time until absorption? Note that this second question is meaningful whether the chain is ergodic or not, while the question of where the chain will go once it leaves the transient states is only of interest in the case of two or more irreducible closed subsets. The reason for this is that if there is only one irreducible closed subset of persistent states, the chain will certainly go there once it leaves the transient states.

The following example illustrates one method of answering these questions.

Example III.4.1. Let $\{X_n\}$ be a Markov chain with state space $S = \{1, 2, \ldots, 9\}$. Let the corresponding transition probability matrix be given by

$$P = \begin{bmatrix} 0 & \frac{2}{3} & \frac{1}{3} & 0 & 0 & 0 & 0 & 0 & 0 \\ \frac{1}{2} & 0 & \frac{1}{2} & 0 & 0 & 0 & 0 & 0 & 0 \\ 1 & 0 & 0 & 0 & 0 & 0 & 0 & 0 & 0 \\ 0 & 0 & 0 & 0 & \frac{1}{2} & \frac{1}{2} & 0 & 0 & 0 \\ 0 & 0 & 0 & \frac{1}{3} & 0 & 0 & \frac{2}{3} & 0 & 0 \\ 0 & 0 & 0 & \frac{2}{3} & 0 & 0 & \frac{1}{3} & 0 & 0 \\ 0 & 0 & 0 & 0 & \frac{1}{4} & \frac{3}{4} & 0 & 0 & 0 \\ \frac{1}{3} & 0 & 0 & 0 & 0 & 0 & \frac{1}{3} & 0 & \frac{1}{3} \\ 0 & \frac{1}{4} & 0 & \frac{1}{2} & 0 & 0 & 0 & \frac{1}{4} & 0 \end{bmatrix}$$

This chain has two irreducible closed sets of persistent states, $C_1 = \{1, 2, 3\}$ and $C_2 = \{4, 5, 6, 7\}$. The periods of these sets are 1 and 2, respectively. States 8 and 9 are transient. Since there are only finitely many transient states, the probability of ultimate absorption is one. (This fact can be verified using Theorem III.3.1. as follows. Consider solutions of the two equations $x_8 = \frac{1}{3} x_9$ and $x_9 = \frac{1}{4} x_8$. The only solution is $x_8 = 0$ and $x_9 = 0$ so this is the maximal solution and hence the probability of absorption is one.) We now go on to find the probability of absorption into C_1 or C_2. These probabilities of absorption usually depend on the transient state in which the chain starts. Therefore consider the probability that a path that starts in state 8 is absorbed into the closed set C_1. This will be done by calculating the absorption probabilities into the individual states of C_1 and then adding these probabilities over the states of C_1. We begin by calculating the probability of being absorbed into state 1 from state 8. Let α_{81} denote this probability. It is easy to see that $p_{81}^{(1)} = \frac{1}{3}$, $p_{81}^{(2)} = 0$, $p_{81}^{(3)} = \frac{1}{3}(1/12)$, and in general $p_{81}^{(2k)} = 0$ and $p_{81}^{(2k+1)} = \frac{1}{3} \cdot (1/12)^k$. Hence $\alpha_{81} = \sum_{k=0}^{\infty} \frac{1}{3}(1/12)^k = \frac{1}{3}(12/11) = 4/11$. In the same way we get $\alpha_{82} = 1/11$ and $\alpha_{83} = 0$. Therefore the probability of absorption into the set C_1 from state 8 is $5/11$. The probabilities $\alpha_{84}, \alpha_{85}, \alpha_{86}, \alpha_{91}, \ldots, \alpha_{96}$ are obtained in the same way.

The expected time until absorption can also be calculated directly. Again there is dependence upon the starting state, so let μ_8 denote the expected time until absorption starting at state 8. The probability of absorption at the first step is $b_1 = \frac{2}{3}$. The probability of absorption at the second step is $b_2 = \frac{1}{3} \cdot \frac{3}{4}$. Continuing in this way we get

$$b_3 = \frac{1}{3} \cdot \frac{1}{4} \cdot \frac{2}{3}$$
$$b_4 = \frac{1}{3} \cdot \frac{1}{4} \cdot \frac{1}{3} \cdot \frac{3}{4},$$
$$\vdots$$
$$b_{2n} = \left(\frac{1}{3}\right)^n \cdot \left(\frac{1}{4}\right)^{n-1} \cdot \frac{3}{4},$$
$$b_{2n+1} = \left(\frac{1}{3}\right)^n \left(\frac{1}{4}\right)^n \cdot \frac{2}{3}$$
$$\vdots$$

Hence

$$\mu_8 = \sum_{k=1}^{\infty} k b_k$$

$$= \sum_{k=1}^{\infty} (2k)\left(\frac{1}{3}\right)^k \left(\frac{1}{4}\right)^{k-1} \cdot \frac{3}{4} + \sum_{k=0}^{\infty} (2k+1)\left(\frac{1}{3}\right)^k \left(\frac{1}{4}\right)^k \cdot \frac{2}{3}$$

$$= 2 \cdot \frac{1}{3} \cdot \frac{3}{4} \sum_{k=1}^{\infty} k(1/12)^{k-1} + 2 \cdot \frac{1}{3} \cdot \frac{1}{4} \cdot \frac{2}{3} \sum_{k=0}^{\infty} k(1/12)^{k-1} + \frac{2}{3} \sum_{k=0}^{\infty} (1/12)^k$$

$$= \frac{1}{2} \cdot (144/121) + \frac{1}{9} \cdot (144/121) + \frac{2}{3}(12/11)$$

$$= \frac{72 + 16 + 88}{121} = \frac{16}{11}.$$

In a similar manner it can be shown that $\mu_9 = 15/11$ (Exercise 1).

The method used in the above example to calculate the absorption probabilities and expected absorption times will be referred to as the *direct approach.* From this example we see that the direct approach method can be rather tedious. Actually for many examples it is almost impossible. The direct approach will be used in the following example so its deficiencies will be clearer.

Example III.4.2. Let $\{X_n\}$ be a Markov chain with transition matrix

$$
P = \begin{bmatrix}
1 & 0 & 0 & 0 \\
0 & 1 & 0 & 0 \\
\frac{1}{3} & 0 & \frac{1}{3} & \frac{1}{3} \\
\frac{1}{4} & \frac{1}{4} & \frac{1}{4} & \frac{1}{4}
\end{bmatrix}.
$$

This chain has two absorbing states, $\{1,2\}$, and two transient states, $\{3,4\}$. Again since the state space is finite the chain will be absorbed from the transient states with probability one. If α_{31} denotes the ultimate absorption probability from state 3 into the closed set $C_1 = \{1\}$, then $\alpha_{31} = \sum_{k=1}^{\infty} p_{31}^{(k)}$. We have $p_{31} = \frac{1}{3}$, $p_{31}^{(2)} = \frac{1}{3} \cdot \frac{1}{3} + \frac{1}{3} \cdot \frac{1}{4}$, and $p_{31}^{(3)} = \frac{1}{3} \cdot \frac{1}{3} \cdot \frac{1}{3} + \frac{1}{3} \cdot \frac{1}{4} \cdot \frac{1}{4} + \frac{1}{3} \cdot \frac{1}{3} \cdot \frac{1}{4} + \frac{1}{3} \cdot \frac{1}{4} \cdot \frac{1}{3}$. It becomes evident that $p_{31}^{(k)}$ will be very difficult to calculate for large k since for the first $k-1$ steps the chain moves freely between states 3 and 4. For the same reason the expected time until absorption is difficult to calculate. At this point we see why Example III.4.1 was easier. This chain did not remain in either of the transient states with positive probability, so if the chain was not absorbed, it moved to the other transient state.

In view of the difficulties encountered in Example III.4.2 we turn to a second method for finding absorption probabilities and mean absorption times. The method will first be discussed for a general Markov chain with a finite state space containing two closed sets and then will be applied to the two above examples. Consider the following transition matrix P in block form:

$$
P = \begin{bmatrix}
P_1 & 0 & 0 \\
0 & P_2 & 0 \\
R_1 & R_2 & Q
\end{bmatrix}.
$$

Let i and j be transient states and let $q_{ij}^{(n)}$ denote the probability of going from i to j in n steps. The previous examples showed that it might be difficult to calculate these probabilities. The first method considered for finding mean absorption times and absorption probabilities required that the $q_{ij}^{(n)}$'s be known. The advantage of the present method is that the individual $q_{ij}^{(n)}$'s are not needed. First consider the question of the mean absorption time from state i.

$$\mu_i = \sum_{k=1}^{\infty} k \cdot P[\text{absorbed from state } i \text{ at step } k]$$

$$= \sum_{k=1}^{\infty} \sum_{m=0}^{k-1} P[\text{absorbed from state } i \text{ at step } k]$$

$$\overset{*}{=} \sum_{m=0}^{\infty} \sum_{k=m+1}^{\infty} P[\text{absorbed from state } i \text{ at step } k]$$

$$= \sum_{m=0}^{\infty} P[\text{absorbed from state } i \text{ sometime after step } m]$$

$$= \sum_{m=0}^{\infty} P[\text{not absorbed from state } i \text{ by step } m]$$

$$= \sum_{m=0}^{\infty} \sigma_i^{(m)} = \sum_{m=0}^{\infty} \sum_{j \in T} q_{ij}^{(m)} \overset{*}{=} \sum_{j \in T} \sum_{m=0}^{\infty} q_{ij}^{(m)}.$$

(*Fubini's theorem applies since all terms are non-negative). Hence we see that it is sufficient to know the *sums*, $\sum_{m=0}^{\infty} q_{ij}^{(m)}$ for all $j \in T$ in order to calculate the mean absorption time from state i. This may not appear to be any better, but if we consider the matrix $N_n = I + Q + Q^2 + \cdots + Q^n$, where I is the identity matrix, we can see that the (i,j)th entry of N_n is $\sum_{m=0}^{n} q_{ij}^{(m)}$. Therefore all that is needed to find the sums $\sum_{m=0}^{\infty} q_{ij}^{(m)}$, and consequently the mean absorption times, is the matrix $N = I + Q + Q^2 + \cdots$. At first glance this does not seem to be any easier than the first method. However, we will show that $N = (I - Q)^{-1}$.

Lemma III.4.1. Let Q be the substochastic matrix corresponding to transitions among the transient states of a finite Markov chain. Then $(I - Q)^{-1} = I + Q + Q^2 + \cdots$ exists.

Proof. We first note that

$$N_n(I - Q) = (I - Q)N_n = I - Q^{n+1}. \tag{4.1}$$

Also

$$N_n = I + Q + \cdots + Q^n = I + \sum_{k=1}^{n} Q^k = \sum_{k=0}^{n} Q^k$$

so N_n converges if and only if $\sum_{k=0}^{n} Q^k$ converges as $n \to \infty$. Consequently if we can show that N_n converges to N, it will follow from (4.1) that

$$N(I-Q) = (I-Q)N = I,$$

that is, $N = (I-Q)^{-1}$.

In order to show that the series $\sum_{k=0}^{n} Q^k$ converges, it suffices to show that each entry of the matrix converges. That is, it suffices to show that for i and j in T, $\sum_{n=0}^{\infty} q_{ij}^{(n)} < \infty$.

First we will show that $\sum_{n=0}^{\infty} q_{ij}^{(n)}$ represents the expected number of visits to state j when the chain starts in state i. Let the chain be denoted by $\{X_n\}_{n=0}^{\infty}$. Assume $X_0 = i$ and define

$$Y_n = 1 \quad \text{if} \quad X_n = j.$$

$$Y_n = 0 \quad \text{if} \quad X_n \neq j.$$

The series $\sum_{n=0}^{\infty} Y_n$ represents the number of visits to state j when the chain starts at state i. Hence $E[\sum_{n=0}^{\infty} Y_n]$ represents the expected number of visits to state j when the chain starts in state i. Using Fubini's theorem we get

$$E\left[\sum_{n=0}^{\infty} Y_n \right] = \sum_{n=0}^{\infty} E(Y_n) = \sum_{n=0}^{\infty} q_{ij}^{(n)}$$

since $E(Y_n) = q_{ij}^{(n)}$.

(Here we are using a more general form of Fubini's theorem, which allows for the interchange of expectation and infinite sum when the integrand is non-negative.)

Another way to represent the expected number of visits to state j when the chain starts in state i is to define a random variable Z_{ij} to be the number of visits to state j when the chain starts in state i. Recall that f_{ij}^* denotes the probability of ever going from i to j and f_{jj}^* denotes the probability of ever returning to state j. Hence

$$P[Z_{ij} = k] = f_{ij}^* (f_{jj}^*)^{k-1} (1 - f_{jj}^*).$$

Using this we get

$$E[Z_{ij}] = \sum_{k=1}^{\infty} k f_{ij}^* (f_{jj}^*)^{k-1} (1 - f_{jj}^*) = \frac{f_{ij}^*}{1 - f_{jj}^*}. \qquad (4.2)$$

Since j is transient we have $f_{jj}^* < 1$ so $E[Z_{ij}] < \infty$. However $E[Z_{ij}]$

$= \sum_{n=0}^{\infty} q_{ij}^{(n)}$, so $\sum_{n=0}^{\infty} q_{ij}^{(n)} < \infty$, that is, $I + Q + Q^2 + \cdots$ converges to a finite matrix, N, and $N = (I - Q)^{-1}$. ▲

The matrix N is often called the fundamental matrix, so this method will be referred to as the *fundamental matrix method*. Using this method, the way to get the expected absorption time from state 8 in Example III.4.1 is to calculate

$$N = (I - Q)^{-1} = \begin{pmatrix} 1 & -\frac{1}{3} \\ -\frac{1}{4} & 1 \end{pmatrix}^{-1} = \begin{pmatrix} 12/11 & 4/11 \\ 3/11 & 12/11 \end{pmatrix}.$$

The expected absorption time from state 8 is the sum of the entries in the first row so $\mu_8 = 12/11 + 4/11 = 16/11$. Similarly $\mu_9 = 15/11$. If $\mathbf{1}'$ denotes a column vector of ones then $N\mathbf{1}'$ is a vector, $\boldsymbol{\mu}'$, in which the ith entry is the mean absorption time from the ith transient state. In the case of Example III.4.2 we get

$$N = \begin{pmatrix} \frac{2}{3} & -\frac{1}{3} \\ -\frac{1}{4} & \frac{3}{4} \end{pmatrix}^{-1} = \begin{pmatrix} \frac{9}{5} & \frac{4}{5} \\ \frac{3}{5} & \frac{8}{5} \end{pmatrix}$$

so $\boldsymbol{\mu}' = N\mathbf{1}' = (13/5, 11/5)'$. In cases where there are many transient states, finding the inverse of $I - Q$ may not be easy. However, it is certainly much easier than the first method discussed which required the calculation of all the $q_{ij}^{(n)}$'s.

We now turn to the question of absorption probabilities from transient state i to persistent state l. For notational convenience assume there are three transient states in the chain. Denote these states by $i, j,$ and k. Let α_{il} denote the probability of being absorbed into state l from state i. Let r_{il} denote the probability of moving from state i to state l in one step. Note that r_{il} is simply the (i, l)th entry of

$$P = \begin{bmatrix} P_1 & 0 & 0 \\ 0 & P_2 & 0 \\ R_1 & R_2 & Q \end{bmatrix}.$$

Similarly use r_{jl} and r_{kl} for the transition probabilities from j and k to l. Let $q_{ij}, q_{ik},$ and so on denote the transition probabilities among the transient states, T. Now

$$\alpha_{il} = r_{il} + q_{ii}r_{il} + q_{ij}r_{jl} + q_{ik}r_{kl} + q_{ii}^{(2)}r_{il} + \cdots$$
$$+ q_{ii}^{(n)}r_{il} + q_{ij}^{(n)}r_{jl} + q_{ik}^{(n)}r_{kl} + \cdots.$$

Since all these terms are non-negative, they can be rearranged to give

$$\alpha_{il} = r_{il} \sum_{n=0}^{\infty} q_{ii}^{(n)} + r_{jl} \sum_{n=0}^{\infty} q_{ij}^{(n)} + r_{kl} \sum_{n=0}^{\infty} q_{ik}^{(n)}.$$

Again we see that the values of the fundamental matrix, N, are sufficient for finding these absorption probabilities. In fact the matrix of all the absorption probabilities is obtained by multiplying N times the matrix R, where R is the matrix consisting of the elements in R_1 and R_2. In particular if P_1 is $m_1 \times m_1$ and P_2 is $m_2 \times m_2$ and Q is $n \times n$ then R_1 is $n \times m_1$ and R_2 is $n \times m_2$. Hence R is $n \times (m_1 + m_2)$ so NR is $n \times (m_1 + m_2)$. The (i,l)th entry of NR is the probability of being absorbed into persistent state l from transient state i.

Returning to Example III.4.1. we have

$$N = \begin{pmatrix} 12/11 & 4/11 \\ 3/11 & 12/11 \end{pmatrix} \quad \text{and} \quad R = \begin{pmatrix} \frac{1}{3} & 0 & 0 & 0 & 0 & 0 & \frac{1}{3} \\ 0 & \frac{1}{4} & 0 & \frac{1}{2} & 0 & 0 & 0 \end{pmatrix}$$

so

$$NR = \begin{pmatrix} \alpha_{81} & \alpha_{82} & \cdots & \alpha_{87} \\ \alpha_{91} & \alpha_{92} & \cdots & \alpha_{97} \end{pmatrix} = \begin{pmatrix} 4/11 & 1/11 & 0 & 2/11 & 0 & 0 & 4/11 \\ 1/11 & 3/11 & 0 & 6/11 & 0 & 0 & 1/11 \end{pmatrix}.$$

From this the probability of absorption into the closed set $C_1 = \{1,2,3\}$ from state 8 is obtained by adding $\alpha_{81} + \alpha_{82} + \alpha_{83} = 5/11$. Probabilities of absorption into other closed sets from the various transient states are handled similarly. Note that when looking for absorption probabilities the technique of reducing the closed sets to an absorbing state, as in (3.1), cannot be used since this operation changes R.

Example III.4.3. Let $\{X_n\}$ be a Markov chain with state space $S = \{1,2,3,4,5,6,7\}$ and transition matrix

$$P = \begin{pmatrix} 1 & 0 & 0 & 0 & 0 & 0 & 0 \\ 0 & \frac{1}{2} & \frac{1}{2} & 0 & 0 & 0 & 0 \\ 0 & \frac{1}{3} & \frac{1}{3} & \frac{1}{3} & 0 & 0 & 0 \\ 0 & 0 & 1 & 0 & 0 & 0 & 0 \\ \frac{1}{2} & 0 & 0 & \frac{1}{4} & \frac{1}{4} & 0 & 0 \\ 0 & 0 & \frac{1}{3} & 0 & \frac{1}{3} & 0 & \frac{1}{3} \\ 0 & 0 & 0 & 0 & 0 & \frac{1}{2} & \frac{1}{2} \end{pmatrix}.$$

In order to find the absorption probabilities and absorption times we begin with

$$Q = \begin{bmatrix} \frac{1}{4} & 0 & 0 \\ \frac{1}{3} & 0 & \frac{1}{3} \\ 0 & \frac{1}{2} & \frac{1}{2} \end{bmatrix}$$

so

$$I - Q = \begin{bmatrix} \frac{3}{4} & 0 & 0 \\ -\frac{1}{3} & 1 & -\frac{1}{3} \\ 0 & -\frac{1}{2} & \frac{1}{2} \end{bmatrix} \quad \text{and} \quad N = (I - Q)^{-1} = \begin{bmatrix} \frac{4}{3} & 0 & 0 \\ \frac{2}{3} & \frac{3}{2} & 1 \\ \frac{2}{3} & \frac{3}{2} & 3 \end{bmatrix}.$$

The mean absorption times from states 5, 6, and 7, respectively, are given by the vector $N\mathbf{1}' = (4/3, 19/6, 31/6)'$. The absorption probabilities from the transient states into the various persistent states are given by

$$NR = \begin{bmatrix} \frac{4}{3} & 0 & 0 \\ \frac{2}{3} & \frac{3}{2} & 1 \\ \frac{2}{3} & \frac{3}{2} & 3 \end{bmatrix} \cdot \begin{bmatrix} \frac{1}{2} & 0 & 0 & \frac{1}{4} \\ 0 & 0 & \frac{1}{3} & 0 \\ 0 & 0 & 0 & 0 \end{bmatrix} = \begin{bmatrix} \frac{2}{3} & 0 & 0 & \frac{1}{3} \\ \frac{1}{3} & 0 & \frac{1}{2} & \frac{1}{6} \\ \frac{1}{3} & 0 & \frac{1}{2} & \frac{1}{6} \end{bmatrix}.$$

In addition to finding the mean of the absorption times to the persistent states, the fundamental matrix can be used to find the second moment of the absorption time. In particular the vector of second moments is given by $\mu^{(2)'} = N(2\mu' - \mathbf{1}')$ where μ is the vector of expected absorption times (Exercise 7). Hence for Example III.4.3 we have $\mu^{(2)'} = (20/9, 166/9, 334/9)'$.

There is one more important question that can be resolved for finite chains with transient states. To illustrate this question, we once again refer to Example III.4.1. This chain is clearly not ergodic since there are two irreducible closed sets. This means that $\lim_{n \to \infty} p_{ij}^{(n)}$ will not be independent of i even though the limit may exist. In particular $\lim_{n \to \infty} p_{i1}^{(n)} = \frac{3}{7}$ for $i = 1, 2, 3$ and $\lim_{n \to \infty} p_{i1}^{(n)} = 0$ for $i = 4, 5, 6, 7$. Now what about $\lim_{n \to \infty} p_{81}^{(n)}$ and $\lim_{n \to \infty} p_{91}^{(n)}$? Intuitively, $\lim_{n \to \infty} p_{81}^{(n)}$ represents the probability that after a long time the chain is in state 1 if it starts in state 8. Once the chain enters C_1 the probability of being in state 1 after a long time is $\frac{3}{7}$. Hence $\lim_{n \to \infty} p_{81}^{(n)}$ should be the product of the probability of entering C_1 from state 8 times $\frac{3}{7}$. This is correct and can be demonstrated as follows:

$$p_{81}^{(n)} = \sum_{k=0}^{n} h_{81}^{(k)} p_{11}^{(n-k)} + \sum_{k=0}^{n} h_{82}^{(k)} p_{21}^{(n-k)}$$

where $h_{8j}^{(k)}$ is the probability of being absorbed into C_1 at state j at time k. (In this example $h_{83}^{(k)} = 0$ for all k.) Using Theorem I.4.4 we get

$$\lim_{n \to \infty} p_{81}^{(n)} = \sum_{k=1}^{\infty} h_{81}^{(k)} \cdot \tfrac{3}{7} + \sum_{k=1}^{\infty} h_{82}^{(k)} \cdot \tfrac{3}{7}$$

$$= (4/11) \cdot (3/7) + (1/11) \cdot (3/7) = (5/11) \cdot (3/7) = 15/77.$$

Similarly

$$\lim_{n \to \infty} p_{82}^{(n)} = 10/77 \quad \text{and} \quad \lim_{n \to \infty} p_{83}^{(n)} = 10/77.$$

The long run probabilities of being in states 1, 2, or 3 starting at state 9 can also be found (Exercise 3).

The situation is somewhat different for $\lim_{n \to \infty} p_{84}^{(n)}$. As above there is no problem in finding the probability of being absorbed into the closed set C_2 from state 8. The problem is that the irreducible subset C_2 does not form an aperiodic chain. Since C_2 has period 2, the limit of $p_{44}^{(n)}$ does not exist as $n \to \infty$. Hence $\lim_{n \to \infty} p_{84}^{(n)}$ does not exist. One could go to subsequences or averages as before, but the values obtained would be of questionable interest.

In summary if from transient state i a chain is absorbed into an irreducible closed set, C, with probability α_i and if C is aperiodic with long run distribution $(\pi_1, \ldots, \pi_m)$, then $\lim_{n \to \infty} p_{ij}^{(n)} = \alpha_i \pi_j$ for $j \in C$. If C is periodic, the limit will not exist.

EXERCISES

1. Show in Example III.4.1 that $\mu_9 = 15/11$ using the direct approach.
2. Find the mean and variance for the absorption times from the transient states for the Markov chain given in Example III.2.5.
3. For Example III.4.1 find $\lim_{n \to \infty} p_{91}^{(n)}$.
4. Reduce the following stochastic matrices to block form and find the absorption probabilities and the mean and variance of the absorp-

tion times. Also find $\lim_{n\to\infty} p_{62}^{(n)}$ and $\lim_{n\to\infty} p_{65}^{(n)}$.

(a) $\quad P = \begin{pmatrix} \frac{1}{2} & 0 & \frac{1}{3} & 0 & \frac{1}{6} & 0 \\ 0 & \frac{1}{3} & 0 & \frac{2}{3} & 0 & 0 \\ 1 & 0 & 0 & 0 & 0 & 0 \\ 0 & 1 & 0 & 0 & 0 & 0 \\ \frac{1}{2} & 0 & 0 & 0 & \frac{1}{2} & 0 \\ \frac{1}{4} & 0 & \frac{1}{4} & \frac{1}{4} & 0 & \frac{1}{4} \end{pmatrix}.$

(b) $\quad P = \begin{pmatrix} \frac{1}{3} & 0 & 0 & 0 & \frac{1}{3} & 0 & \frac{1}{3} & 0 \\ 0 & 0 & 0 & 0 & 0 & 0 & 0 & 1 \\ \frac{1}{4} & 0 & 0 & \frac{1}{2} & 0 & \frac{1}{4} & 0 & 0 \\ 0 & 1 & 0 & 0 & 0 & 0 & 0 & 0 \\ \frac{1}{2} & 0 & 0 & 0 & 0 & 0 & \frac{1}{2} & 0 \\ 0 & 0 & 0 & 1 & 0 & 0 & 0 & 0 \\ 0 & 0 & 0 & 0 & 1 & 0 & 0 & 0 \\ 0 & \frac{1}{2} & 0 & 0 & 0 & 0 & 0 & \frac{1}{2} \end{pmatrix}.$

5.   Show that $\sum_{k=1}^{\infty} k p^k = p/(1-p)^2$ if $0 < p < 1$, thus justifying Equation 4.2.

6.   Let T be a set of m transient states. Let Γ be the set of all paths of length $m + 1$ which begin at state $i \epsilon T$. Show that at least one path in Γ must leave T.

†7.   Show that $\mu^{(2)\prime} = N(2\mu' - 1')$, where $\mu^{(2)\prime}$ is the column vector of second moments and $1'$ is a column vector of ones.

†8.   An alternative approach to finding mean absorption times for a stochastic matrix P is the following. Assume that P has the form

$$P = \left(\begin{array}{c|c} P_1 & 0 \\ \hline R & Q \end{array} \right).$$

(See Section III.2.) For the matrix $M = (m_{ij})$ of expected times until absorption from transient state i to persistent state j consider $\sum_{k=1}^{\infty} k Q^{k-1} R$. Show how this can be used to get the result $\mu' = N \cdot 1'$.

SECTION 5   **RATIO LIMIT THEOREMS**

The investigation of $\lim_{n\to\infty} p_{ij}^{(n)}$ is now complete. In the positive persistent, irreducible, aperiodic case the limit is the unique invariant probability distribution, π, corresponding to the transition matrix, P. In the case of irreducible positive persistent periodic chains the $\lim_{n\to\infty} p_{ij}^{(n)}$ may not exist but $(1/N)\cdot\sum_{n=1}^{N} p_{ij}^{(n)}$ does converge to $1/\mu_j$ as $N\to\infty$ and the vector $(1/\mu_1, 1/\mu_2, \cdots)$ is an invariant probability distribution for P. In the case of an irreducible null persistent chain the limit of $p_{ij}^{(n)}$ is zero for all i and j, so no invariant probability distribution can be found using the above limits. In spite of the fact that for an irreducible null persistent chain there is no invariant *probability* distribution, there is an invariant distribution. The existence of this invariant distribution will be established by considering a generalization of the limit of averages, $(1/N)\sum_{k=1}^{N} p_{ij}^{(k)}$. Namely, we will consider

$$\lim_{N\to\infty} \frac{\displaystyle\sum_{n=0}^{N} p_{ij}^{(n)}}{\displaystyle\sum_{n=0}^{N} p_{kl}^{(n)}}.$$

Definition III.5.1.   *Let $g_{ij}^{(n)}$ denote the probability of visiting state j at time n without having returned to state i before time n. That is*

$$g_{ij}^{(n)} = P[X_n = j, X_{n-1} \neq i, \ldots X_1 \neq i | X_0 = i].$$

Let $g_{ij}^* = \sum_{n=1}^{\infty} g_{ij}^{(n)}$.

We will show that g_{ij}^* represents the expected number of visits to state j before returning to state i. To see this let Y_n be a random variable which is one if $X_n = j$, $X_{n-1} \neq i$, $X_{n-2} \neq i, \ldots$, $X_1 \neq i$, $X_0 = i$ and is zero otherwise. The total number of visits to j before returning to i is given by $\sum_{n=1}^{\infty} Y_n$, so the expected number of visits is

$$E\left(\sum_{n=1}^{\infty} Y_n\right) = \sum_{n=1}^{\infty} E(Y_n) = \sum_{n=1}^{\infty} g_{ij}^{(n)}.$$

The interchange of expectation and summation is justified by Fubini's theorem since $Y_n \geq 0$. (The form of Fubini's theorem needed here is more general than Theorem I.4.2.) The quantity g_{ij}^* will be used throughout the remainder of this section.

We now state the Doeblin ratio theorem, which will yield a way of finding an invariant distribution for null persistent chains.

Theorem III.5.1.

Let i,j,k,l be any states of an irreducible persistent chain. Then

$$\lim_{N \to \infty} \frac{\displaystyle\sum_{n=0}^{N} p_{ij}^{(n)}}{\displaystyle\sum_{n=0}^{N} p_{kl}^{(n)}} .$$

Before proving this theorem some discussion of its meaning will be given. First note that if the irreducible chain happens to be positive persistent, the theorem contains no new information. That is in the irreducible positive persistent case we already know that $[1/(N+1)]\sum_{n=0}^{N} p_{ij}^{(n)}$ converges to $1/\mu_j > 0$. Hence if the numerator and denominator of the ratio in Doeblin's theorem are divided by $N+1$, the ratio converges to μ_l/μ_j. In this case we not only establish the existence of the limit but also obtain its value. In view of the fact that the existence of the limit in the positive persistent case is obvious, the importance of the Doeblin ratio theorem lies in its application to the null persistent case. In the null persistent case if the numerator and denominator are divided by $N+1$ the ratio goes to the indeterminate form of $0/0$. The fact that the limit actually exists is now an interesting result.

Proof of Doeblin's ratio theorem. The basic idea of this proof is to first consider two special cases and then deduce the general case from these. For Case 1 we let $k=l=j \neq i$. By (2.2) of Chapter II, we have that

$$\sum_{n=0}^{N} p_{ij}^{(n)} = \sum_{n=0}^{N} \sum_{k=0}^{n} f_{ij}^{(n-k)} p_{jj}^{(k)}. \tag{5.1}$$

Using Fubini's theorem to justify the interchange of the order of summation, we get

$$\sum_{n=0}^{N} p_{ij}^{(n)} = \sum_{k=0}^{N} \sum_{n=k}^{N} f_{ij}^{(n-k)} p_{jj}^{(k)} = \sum_{k=0}^{N} p_{jj}^{(k)} \sum_{m=0}^{N-k} f_{ij}^{(m)}.$$

Dividing both sides by $\sum_{n=0}^{N} p_{jj}^{(n)}$ we get

$$\frac{\displaystyle\sum_{n=0}^{N} p_{ij}^{(n)}}{\displaystyle\sum_{n=0}^{N} p_{jj}^{(n)}} = \frac{\displaystyle\sum_{k=0}^{N} p_{jj}^{(k)} \sum_{m=0}^{N-k} f_{ij}^{(m)}}{\displaystyle\sum_{n=0}^{N} p_{jj}^{(n)}} .$$

Since all of the states are persistent, we have $\sum_{n=0}^{\infty} p_{jj}^{(n)} = \infty$ so

$$\frac{p_{jj}^{(n)}}{\displaystyle\sum_{k=0}^{n} p_{jj}^{(k)}} \to 0 \quad \text{as } n \to \infty.$$

Hence by Exercise 13 of Section I.4 we get

$$\frac{\displaystyle\sum_{n=0}^{N} p_{ij}^{(n)}}{\displaystyle\sum_{n=0}^{N} p_{jj}^{(n)}} \to \sum_{m=0}^{\infty} f_{ij}^{(m)} = f_{ij}^{*} = 1$$

since the chain is irreducible and persistent.

For Case 2 we let $k = l = i$. In Case 1, $p_{ij}^{(n)}$ was expressed as a sum of probabilities according to when the chain made its *first* visit to j from i. In this case $p_{ij}^{(n)}$ will be expressed as a sum of probabilities according to when the chain made its *last* return to i before going to j. Using Fubini's theorem again we get

$$\sum_{n=0}^{N} p_{ij}^{(n)} = \sum_{n=0}^{N} \sum_{k=0}^{n} p_{ii}^{(k)} g_{ij}^{(n-k)}$$

$$= \sum_{k=0}^{N} p_{ii}^{(k)} \sum_{n=k}^{N} g_{ij}^{(n-k)} = \sum_{k=0}^{N} p_{ii}^{(k)} \sum_{m=0}^{N-k} g_{ij}^{(m)}.$$

Dividing both sides by $\sum_{n=0}^{N} p_{ii}^{(n)}$, we get

$$\frac{\displaystyle\sum_{n=0}^{N} p_{ij}^{(n)}}{\displaystyle\sum_{n=0}^{N} p_{ii}^{(n)}} = \frac{\displaystyle\sum_{k=0}^{N} p_{ii}^{(k)} \sum_{m=0}^{N-k} g_{ij}^{(m)}}{\displaystyle\sum_{n=0}^{N} p_{ii}^{(n)}} .$$

Taking the limit as $N \to \infty$ and applying the same exercise as above, we see that the limit is g_{ij}^*. The general case is obtained by taking products of ratios of the form of Cases 1 and 2 and then using the fact that the limit of a product is the product of the limits when the limits exist. In particular,

$$
\frac{\displaystyle\sum_{n=0}^{N} p_{ij}^{(n)}}{\displaystyle\sum_{n=0}^{N} p_{kl}^{(n)}} = \frac{\displaystyle\sum_{n=0}^{N} p_{ij}^{(n)}}{\displaystyle\sum_{n=0}^{N} p_{jj}^{(n)}} \cdot \frac{\displaystyle\sum_{n=0}^{N} p_{jj}^{(n)}}{\displaystyle\sum_{n=0}^{N} p_{jl}^{(n)}} \cdot \frac{\displaystyle\sum_{n=0}^{N} p_{jl}^{(n)}}{\displaystyle\sum_{n=0}^{N} p_{ll}^{(n)}} \cdot \frac{\displaystyle\sum_{n=0}^{N} p_{ll}^{(n)}}{\displaystyle\sum_{n=0}^{N} p_{kl}^{(n)}} .
$$

Taking the limit of both sides of this equation we get

$$
\lim_{N \to \infty} \frac{\displaystyle\sum_{n=0}^{N} p_{ij}^{(n)}}{\displaystyle\sum_{n=0}^{N} p_{kl}^{(n)}} = f_{ij}^* \cdot \frac{1}{g_{jl}^*} \cdot f_{jl}^* \cdot \frac{1}{f_{kl}^*} = \frac{1}{g_{jl}^*} .
$$

Note that $g_{jl}^* \neq 0$ since it represents the expected number of visits to state l without returning to j. ▲

There are different products of ratios that could be used to prove the general case from Cases 1 and 2. Under some of these other approaches the limit may have any of the following equivalent forms (Exercise 6):

$$
\frac{g_{ij}^*}{g_{il}^*} = \frac{g_{kj}^*}{g_{kl}^*} = \frac{1}{g_{jl}^*} = \frac{g_{jj}^*}{g_{jl}^*} .
$$

The following theorem shows that every irreducible persistent Markov chain does have an invariant distribution although it is not necessarily a finite distribution. This distribution is unique up to multiplicative constants.

Theorem III.5.2.

In the case of an irreducible persistent Markov chain, the system of equations

$$
\alpha_j = \sum_{k=1}^{\infty} \alpha_k p_{kj} \tag{5.2}
$$

is solved by $\alpha_j = g_{ij}^*$ where i is arbitrary. Any two non-negative nonzero solutions of this equation differ only by a multiplicative constant.

Proof. Consider

$$\sum_{j=1}^{\infty} g_{ij}^* p_{jk} = \sum_{j=1}^{\infty} \sum_{n=1}^{\infty} g_{ij}^{(n)} p_{jk} = \sum_{n=1}^{\infty} \sum_{j=1}^{\infty} g_{ij}^{(n)} p_{jk}.$$

Recall that $g_{ij}^{(n)} = P$[visit state j at time n without returning to state i]. Hence $\sum_{j \neq i} g_{ij}^{(n)} p_{jk}$ represents the probability of visiting state k at time $n+1$ without returning to i. When $j = i$ the quantity $g_{ii}^{(n)}$ is equal to $f_{ii}^{(n)}$ since both of them represent the probability of returning to i for the first time at time n. Because of this we have for any persistent chain that $\sum_{n=1}^{\infty} g_{ii}^{(n)} = 1$. Also note that $p_{ij} = g_{ij}^{(1)}$.

In view of the above discussion we simplify $\sum_{n=1}^{\infty} \sum_{j=1}^{\infty} g_{ij}^{(n)} p_{jk}$ as follows:

$$\sum_{n=1}^{\infty} \sum_{j=1}^{\infty} g_{ij}^{(n)} p_{jk} = \sum_{n=1}^{\infty} g_{ik}^{(n+1)} + p_{ik} \sum_{n=1}^{\infty} g_{ii}^{(n)}$$

$$= \sum_{n=1}^{\infty} g_{ik}^{(n+1)} + p_{ik} = g_{ik}^*.$$

Hence $\{ g_{ij}^* \}_{j=1}^{\infty}$ satisfies the system of equations (5.2).

In order to show that any two non-negative, nonzero solutions to (5.2) differ by a constant, let $\{ \beta_j \}$ be such a solution. Since the solution is nonzero, there exists a state k such that $\beta_k > 0$. Next note that since $\beta_j = \sum_{i=1}^{\infty} \beta_i p_{ij}$, it follows that $\beta_j = \sum_{i=1}^{\infty} \beta_i p_{ij}^{(n)}$ for all $n \geq 1$. Further, by irreducibility we know that there is some n_0 such that $p_{kj}^{(n_0)}$ is strictly positive, hence $\beta_j = \sum_{i=1}^{\infty} \beta_i p_{ij}^{(n_0)} \geq \beta_k p_{kj}^{(n_0)} > 0$. That is, a non-negative, non-zero solution to (5.2) must in fact be a strictly positive solution.

We know that $\alpha_j = g_{ij}^*$ is a positive solution to (5.2). Let $\{ \beta_j \}$ by any other positive solution. Define $r_{ij} = (\beta_j / \beta_i) p_{ji}$. It can be shown that $\{ r_{ij} \}$ is a transition probability matrix for a persistent irreducible chain (Exercise 3). It follows from Cases 1 and 2 of Theorem III.5.1 that

$$1 \leftarrow \frac{\displaystyle\sum_{n=1}^{N} r_{ji}^{(n)}}{\displaystyle\sum_{n=1}^{N} r_{ii}^{(n)}} = \frac{\beta_i}{\beta_j} \frac{\displaystyle\sum_{n=1}^{N} p_{ij}^{(n)}}{\displaystyle\sum_{n=1}^{N} p_{ii}^{(n)}} \to \frac{\beta_i}{\beta_j} g_{ij}^*.$$

Hence $\beta_j = \beta_i g_{ij}^* = \beta_i \alpha_j$, $j = 1, 2, \ldots$ so β_j is a constant multiple of α_j.   ▲

We summarize this theorem by saying that any irreducible persistent Markov chain has an invariant distribution. The positive persistent chains have an invariant *probability* distribution. In the case of null persistent chains the g_{ij}^*'s form an infinite distribution on S, that is, $\sum_{j=1}^{\infty} g_{ij}^* = \infty$ (Exercise 4).

One can go further in the study of limit theorems for various cases of Markov chains. For example, Orey (1961) has studied the existence of $\lim_{n \to \infty} (p_{ij}^{(n)} / p_{kl}^{(n+m)})$ for certain Markov chains. As in Doeblin's theorem the interesting case is the null persistent case in which the ratio goes to an indeterminate form. Chains for which this limit exists are said to have the strong ratio limit property. Rather than pursuing such ratio limit theorems here, we will return in Chapter IV to the basic question of analyzing a discrete–time, discrete-space Markov chain.

In the final section of this chapter some applications of Markov chains to various fields are given. In particular it is shown how Markov chains do arise naturally in applications and how an analysis of the chain as discussed in this chapter provides useful information.

EXERCISES

1.   Using the Markov chain defined in Exercise III.2.4 and Doeblin's ratio limit theorem, find

$$\lim_{N \to \infty} \left\{ \sum_{n=0}^{N} p_{ij}^{(n)} / \sum_{n=0}^{N} p_{kl}^{(n)} \right\}$$

     when
     (a) $j = 1$ and $l = 2$
     (b) $j = 2$ and $l = 3$.

*2.   Using the Markov chain defined in Exercise III.2.5, calculate g_{1j}^*, $j = 1, 2, 3, \ldots$, and verify that $\{ g_{1j}^* \}$ is an invariant measure for P.

3.   Show that $R = (r_{ij})$ as given in Theorem III.5.2. is a transition probability matrix for a persistent irreducible chain.

4.   Prove that $\sum_{j=1}^{\infty} g_{ij}^* = \infty$ in the null persistent case.

*5.   Prove that $\sum_{j=1}^{\infty} g_{ij}^* < \infty$ in the positive persistent case.

6.   Show how the different forms of the limit in Doeblin's ratio theorem can be obtained.

7.   In Chapter II it was noted that $p_{ij}^{(n)}$ can be written in the equivalent

forms

$$\sum_{k=0}^{n} f_{ij}^{(k)} p_{jj}^{(n-k)} = \sum_{k=0}^{n} f_{ij}^{(n-k)} p_{jj}^{(k)}$$

when $i \neq j$. It was also stated that in some instances one form is easier to use than the other. As an example of this, prove Case 1 of Theorem III.5.1 by using in (5.1) the form $p_{ij}^{(n)} = \sum_{k=0}^{n} f_{ij}^{(k)} p_{jj}^{(n-k)}$. To some readers this approach may seem easier than the one given and to others it may seem more difficult.

8. Find a stochastic matrix P such that:
    (a) P has an invariant probability distribution but no long run distribution.
    (b) P has an invariant distribution but no invariant probability distribution.
    (c) the invariant probability distribution for P is not unique.

9. Let $\{X_n\}$ be a Markov chain with transition matrix

$$P = \begin{bmatrix} \frac{1}{2} & 0 & \frac{1}{3} & 0 & \frac{1}{6} \\ \frac{1}{4} & \frac{1}{4} & 0 & \frac{1}{2} & 0 \\ 1 & 0 & 0 & 0 & 0 \\ 0 & \frac{1}{3} & \frac{1}{3} & \frac{1}{3} & 0 \\ \frac{1}{4} & 0 & \frac{1}{2} & 0 & \frac{1}{4} \end{bmatrix}$$

Let the state space be $S = \{1,2,3,4,5\}$.
(a) Find the mean recurrence time to state 1.
(b) Find the expected number of visits to state 1 before returning to state 5.
(c) Find the mean absorption time from state 2.
(d) Find $\lim_{n \to \infty} P^n$.

SECTION 6 APPLICATIONS

In this section we will present some applications of Markov chains to a variety of fields. The purpose of this presentation is to illustrate the use of some of the theoretical work presented to this point and to give the reader some appreciation for the importance and wide applicability of Markov chains to practical problems.

These applications have been found in various technical journals. In most cases we will describe the basic experiment or problem and discuss those portions of the articles which relate to material presented earlier in this chapter. We will describe what *has* been done as best we can, but we do not claim that it is what *should* have been done. In fact there are instances where we doubt if the analysis given is correct. We have included examples like this because the reader may well be confronted with this situation. For each application the reader should consider the following questions:

1. Are the states chosen reasonably?
2. Does the Markov property hold?
3. Are the transition probabilities determined in a reasonable way?
4. Is the process stationary?

These and other questions should be answered to determine whether or not it is really appropriate to study the process as a stationary Markov chain. At this point we would encourage the reader to be rather critical and perhaps even skeptical about the procedures used. However, the reader should keep in mind that in many applications a researcher is looking for a model to approximate a real-world situation. The "proof of the pudding" is not how well certain mathematical assumptions are satisfied, but how useful the model is to workers in an applied field. We are not in a position to make this final judgment.

Application III.6.1. "Projection of Farm Numbers for North Dakota with Markov Chains," by Ronald D. Krenz, *Agricultural Economics Research*, **XVI**, No. 3, p. 77, July 1964.

The author was interested in studying changes in the distribution of the sizes of farms in North Dakota. The states of the chain were defined in terms of farm sizes as follows:

State	S_0	S_1	S_2	S_3	S_4	S_5	S_6
Farm size (in acres)	0–9	10–99	100–179	180–259	260–499	500–999	1000 or more

Note that realistically a "farm" falling in state S_0 is not actually a farm.

Since for this problem there was no mathematical way to determine the transition matrix P, the transition probabilities were estimated by using census data. In this case U.S. census data was used for five-year periods from 1935 to 1960. To further assist in the construction of P, the following assumptions were made:

1. Any farm reaching state S_6 would remain there forever.

2. Any farm that increased in size would go to the next higher level; that is, $p_{ij} = 0$ if $j - i \geqslant 2$.
3. Any farm that decreased in size would go to S_0 (and hence could no longer be a considered a farm).

One reason for making assumptions like these is that the census data did not contain information as to what happended to a particular farm but only gave the total number of farms of various sizes. The totals for the years 1935 and 1940 are given as follows:

	S_1	S_2	S_3	S_4	S_5	S_6
1935	3,948	13,572	5,552	35,133	19,891	5,250
1940	2,985	10,415	4,491	29,620	19,371	6,405

From these data, using the assumptions made above, we see that to obtain 6405 farms in S_6 in 1940 we must add 1155 farms from those in S_5 to the 5250 which were in S_6 in 1935. This leaves 18,736 farms in S_5, so a total of 635 farms must have moved from S_4 in 1935 to S_5 in 1940. The results of similar calculations for the remaining states are summarized in Table III.6.1.

Table III.6.1.   Numbers of Farms in S_i in 1935 and in S_j in 1940 using assumptions 1–3.

		Sizes of groups in 1940							Row
		S_0	S_1	S_2	S_3	S_4	S_5	S_6	Total
	S_0	0	0	0	0	0	0	0	0
	S_1	963	2,985	0	0	0	0	0	3,948
Sizes of	S_2	3,157	0	10,415	0	0	0	0	13,572
groups	S_3	1,061	0	0	4,491	0	0	0	5,552
in 1935	S_4	4,878	0	0	0	29,620	635	0	35,133
	S_5	0	0	0	0	0	18,736	1,155	19,891
	S_6	0	0	0	0	0	0	5,250	5,250
Column total		10,059	2,985	10,415	4,491	29,620	19,371	6,405	

Assuming there was nothing special about the era from 1935 to 1940, such a table could be constructed for the other five-year periods and could be used to find a matrix P by first adding the tables together elementwise (i.e., using matrix addition) and then dividing each row by the corresponding sum of row totals. This was done with the resulting stochastic matrix being used as the probability transition matrix in this study. This matrix is given below.

$$P = \begin{bmatrix} 1 & 0 & 0 & 0 & 0 & 0 & 0 \\ .196 & .804 & 0 & 0 & 0 & 0 & 0 \\ .255 & 0 & .745 & 0 & 0 & 0 & 0 \\ .131 & 0 & 0 & .830 & .039 & 0 & 0 \\ .010 & 0 & 0 & 0 & .849 & .051 & 0 \\ 0 & 0 & 0 & 0 & 0 & .942 & .058 \\ 0 & 0 & 0 & 0 & 0 & 0 & 1 \end{bmatrix}$$

Note that for this matrix the ith row and column correspond to state S_{i-1}.

It is easy to see that the Markov chain corresponding to P is nonergodic because the states S_0 and S_6 are absorbing states. In this case the absorption probabilities and expected absorption times may be found. These values were found using the fundamental matrix approach and the results are given below.

		S_0	S_6
	S_1	1	0
Transient	S_2	1	0
State	S_3	.922	.078
	S_4	.661	.339
	S_5	0	1

Absorption Probabilities

	S_1	5.11
Transient	S_2	3.92
State	S_3	8.77
	S_4	12.48
	S_5	17.24

Expected Absorption
Times

Here the expected absorption times are given in terms of a time unit of 5 years. To change these values to the expected number of years until absorption, multiply the values by 5.

The transition matrix P was also used to predict the number of farms of various sizes in the future. The author first considered 1935 as a base year and used P to estimate the distribution of farm sizes in 1960. That is, the vector representing the distribution of farm sizes in 1935 was multiplied by P^5 to give the distribution of farm sizes in 1960. The resulting vector was quite close to the actual vector for 1960. This same approach was used to predict the distribution of farm sizes in 1975 and 2000. These predictions turn out to be dependent on the choice of base year, and so the reliability of such predictions is questionable. As one might guess, the projections indicate that in the future most of the farms in North Dakota will consist of 1000 acres or more.

Application III.6.2. "A Markov Chain Analysis of Caries Process with Consideration for the Effect of Restoration", by K. H. Lee, *Archives of Oral Biology*, **13**, pp. 1119–1132; September 1968. (Permission to reproduce the transition matrices was granted by Pergamon Press, Ltd.).

In this study a total of 1080 dental records on the maxillary second premolars of 184 grade and high school pupils observed in six-month intervals over 3 years by a single examiner were used. Five surfaces of the tooth were considered. These surfaces were labeled as follows:

X_1 = occlusal surface
X_2 = mesial surface
X_3 = distal surface
X_4 = lingual surface
X_5 = buccal surface.

Each of these surfaces was classified as being in one of three conditions. The conditions were coded as follows:

Healthy surface = 0
Decayed surface = 1
Restored surface = 2.

Using this notation, the condition of a tooth can be described by a 5-tuple of zeros, ones, and twos corresponding to X_1 through X_5. For example, the 5-tuple $(0,0,0,0,0)$ represents a healthy tooth, $(1,1,1,1,1)$ represents a missing tooth, and $(2,0,1,0,0)$ represents a tooth with occlusal restoration, a decayed lesion on the distal surface, and healthy mesial, lingual, and buccal surfaces.

If each possible 5-tuple were considered as a state in a Markov chain, there would be $3^5 = 243$ possible states. However, many of the combinations were not considered in the analysis. In fact, only 18 of the 243 possible states were considered. The author analyzed the data two different ways. In the first analysis the two situations of "loss of tooth," denoted by $(1,1,1,1,1)$, and "restoration" were considered as absorbing states. The "restoration" state was defined by grouping together all states (5-tuples) containing at least one 2. For this analysis, the following transition matrix was obtained from the data. The ten states used were $(0,0,0,0,0)$, $(0,0,1,0,0)$, $(0,1,0,0,0)$, $(0,1,1,0,0)$, $(1,0,0,0,0)$, $(1,0,1,0,0)$, $(1,1,0,0,0)$, $(1,1,1,0,0)$, "restoration," and $(1,1,1,1,1)$ and are represented in the matrix P in this order:

$$P = \begin{bmatrix}
.849 & .070 & .034 & .011 & .018 & .003 & 0 & .003 & .012 & 0 \\
0 & .747 & 0 & .187 & 0 & .022 & 0 & 0 & .044 & 0 \\
0 & 0 & .703 & .266 & 0 & 0 & .031 & 0 & 0 & 0 \\
0 & 0 & 0 & .905 & 0 & 0 & 0 & .037 & .058 & 0 \\
0 & 0 & 0 & 0 & .714 & .179 & 0 & .036 & .036 & .035 \\
0 & 0 & 0 & 0 & 0 & .750 & 0 & .250 & 0 & 0 \\
0 & 0 & 0 & 0 & 0 & 0 & .500 & .500 & 0 & 0 \\
0 & 0 & 0 & 0 & 0 & 0 & 0 & .900 & .050 & .050 \\
0 & 0 & 0 & 0 & 0 & 0 & 0 & 0 & 1 & 0 \\
0 & 0 & 0 & 0 & 0 & 0 & 0 & 0 & 0 & 1
\end{bmatrix}.$$

The purpose in considering this chain was to analyze the progression of the decay process with the "missing tooth" state and the "restoration" state as absorbing states. Since this chain is reducible the absorption probabilities and the expected time until absorption were calculated.

In the second analysis, Dr. Lee considered the closed set consisting of the restored teeth. In this case only nine states were actually observed in the data. In the transition matrix given below the states are $(2,0,0,0,0)$, $(2,0,1,0,0)$, $(2,0,2,0,0)$, $(2,1,0,0,0)$, $(2,1,1,0,0)$, $(2,2,1,0,0)$, $(2,2,2,0,0)$, and $(1,1,2,0,0)$, respectively:

$$P = \begin{bmatrix}
.875 & 0 & 0 & .125 & 0 & 0 & 0 & 0 & 0 \\
0 & .143 & .714 & 0 & 0 & .143 & 0 & 0 & 0 \\
0 & .072 & .714 & 0 & 0 & .214 & 0 & 0 & 0 \\
0 & 0 & 0 & .500 & .333 & 0 & 0 & .167 & 0 \\
0 & 0 & 0 & 0 & .727 & .091 & .182 & 0 & 0 \\
0 & 0 & 0 & 0 & .118 & .823 & 0 & .059 & 0 \\
0 & 0 & 0 & 0 & .100 & 0 & .850 & .050 & 0 \\
0 & 0 & 0 & 0 & 0 & 0 & .050 & .950 & 0 \\
0 & 0 & 0 & 0 & 0 & 1 & 0 & 0 & 0
\end{bmatrix}.$$

The purpose in considering this chain was to study the decay behavior under the influence of restoration. Since this chain is ergodic, the long run distribution was found.

From the analysis of these two situations, many interesting observations were given by Dr. Lee. The following are a few of these observations:

i.) From states $(0,1,0,0,0)$, $(1,0,1,0,0)$, and $(1,1,0,0,0)$ the probability of going directly into the restoration state is zero. This indicates that either patients do not seek help for these problems or dentists do not treat these problems.

ii.) In the second analysis, the states $(2,1,1,0,0)$, $(2,1,2,0,0)$, $(2,2,1,0,0)$, and $(2,2,2,0,0)$ form an irreducible closed subset with limiting probabilities .16, .08, .33, and .43, respectively.

iii.) The effect of restoration was demonstrated in this paper. For example, the time to remain "as is" for an occlusal decay, $(1,0,0,0,0)$, is about 1.75 years as compared to 4 years for the corresponding restoration, $(2,0,0,0,0)$.

Application III.6.3. "Markov Chain Analysis in Geography: An application to the Movement of Rental Housing Areas," by W. A. V. Clark, *Annals of the Association of American Geographers*, **55**, pp. 351–359, 1965. (Tables 6.2 and 6.3 reproduced by permission of the Association of American Geographers.)

The basic question considered in this paper is that of the average rents of rental properties in central city tracts between 1940 and 1960. For each tract in a central city, the average rent for rental property was calculated and placed into one of the states defined as follows:

S_1	S_2	S_3	S_4	S_5	S_6	S_7	S_8	S_9	S_{10}	S_{11}
\$0–10	10–20	20–30	30–40	40–50	50–60	60–70	70–80	80–90	90–100	>100

These averages were calculated for the years 1940, 1950, and 1960. A transition probability matrix corresponding to the period 1940–1950 was found by estimating p_{ij} using

$$\frac{\text{number of tracts in state } i \text{ in 1940 and in state } j \text{ in 1950.}}{\text{number of tracts in state } i \text{ in 1940}}$$

As in Application III.6.1, the data from 1940 to 1950 and from 1950 to 1960 were combined to give a single probability transition matrix as follows: Define

$a_{ij} = $ number of tracts in state i in 1940 and in state j in 1950

and define b_{ij} similarly for the period from 1950 to 1960. Then the elements of P were estimated by

$$p_{ij} = \frac{a_{ij} + b_{ij}}{\sum_j a_{ij} + \sum_j b_{ij}}.$$

Since these estimates of p_{ij} would be quite unreliable if there were only a few tracts, the author concentrated on larger cities such as Detroit, Pittsburgh, Indianapolis, and St. Louis. First, Detroit and Pittsburgh were studied with rents undeflated. Then Detroit, Pittsburgh, Indianapolis, and St. Louis were examined with deflated rents. The probability transition matrices for Detroit for undeflated and deflated rents are given in Tables III.6.2 and III.6.3, respectively.

Table III.6.2. Transition probability matrix for average contract rents by tract for Detroit.

	0–10	10–20	20–30	30–40	40–50	50–60	60–70	70–80	80–90	90–100	>100
0–10	0	0	0	0	.157	.157	.211	.211	.211	.053	0
10–20	0	0	.461	.461	.077	0	0	0	0	0	0
20–30	0	0	.029	.647	.294	.012	.012	0	0	0	0
30–40	0	0	0	.024	.488	.358	.106	.024	0	0	0
40–50	0	0	0	0	.061	.297	.405	.174	.056	.004	.004
50–60	0	0	0	0	0	.029	.088	.485	.265	.103	.029
60–70	0	0	0	0	0	0	0	.190	.333	.286	.190
70–80	0	0	0	0	0	0	0	.111	.444	.222	.222
80–90	0	0	0	0	0	0	0	0	0	.429	.571
90–100	0	0	0	0	0	0	0	0	0	.200	.800
>100	0	0	0	0	0	0	0	0	0	.500	.500

Table III.6.3. Transition probability matrix for average contract rents (deflated) by tract for Detroit.

0	0	0	0	.158	.105	.158	.158	.158	.158	.105
0	0	.750	.125	.125	0	0	0	0	0	0
0	0	.204	.680	.097	.009	0	0	0	0	.009
0	0	0	.295	.585	.101	.014	.005	0	0	0
0	0	0	.022	.466	.422	.067	.018	0	.004	0
0	0	0	.007	.061	.458	.336	.084	.046	.007	0
0	0	0	0	.026	.079	.474	.211	.184	.026	0
0	0	0	0	0	0	.200	.400	.266	.133	0
0	0	0	0	0	0	.100	.500	.200	.200	0
0	0	0	0	0	0	0	0	.375	.250	.375
0	0	0	0	0	0	0	0	.125	.250	.625

By comparing the transition probability matrices for the deflated rents of these four cities, the following trends were observed:

i)   there was a greater tendency for Detroit tracts to move to a lower average contract rent than any of the other three cities. This tendency can be seen by the fact that there is a larger number of positive probabilities below the main diagonal for Detroit.

ii)   There was a greater tendency for Detroit and Pittsburgh tracts to maintain the same average contract rent than for St. Louis and Indianapolis. This tendency can be seen by comparing the magnitudes of the probabilities on the main diagonal.

iii)   There was a strong tendency in all cities for the average contract rent to move up one class above the starting state.

Application III.6.4. "Monopoly as a Markov Process," by Robert Ash and Richard Bishop, *Mathematics Magazine*, **45**, pp. 26–29, January 1972. (Tables III.6.4 and III.6.5 are reproduced with the permission of the Mathematical Association of America.)

In this article, the game of Monopoly was studied as a Markov chain. By using the appropriate transition matrices, the authors were able to calculate

   i)   the long run probabilities of landing on each of the properties,
  ii)   the expected income from the bank on each turn, and
 iii)   the expected rent per opponents' turn.

These quantities can be used to develop strategies for playing Monopoly.

In trying to determine the state space and corresponding transition matrix it would be natural to suppose that each of the 40 positions of the board should be a state of the chain with the transition probabilities determined by the throw of two fair dice. However, further consideration of the rules of the game shows that this simple approach would not work. In any Markov chain, the probability of going from state i to state j is determined solely by chance, that is, free-will choices are forbidden. However, in Monopoly a person in jail has a choice of paying his way out of jail or staying in jail until his third turn or until he throws doubles. In order to resolve this difficulty, the authors assumed that one of two strategies would always be followed and that all players would follow the same strategy:

1.   The remain-in-jail (RIJ) rule. The player stays in jail until he throws doubles or until his third turn, whichever comes first.

2.   The leave-jail (LJ) rule. The player always exits on the first turn, either by paying his way out or by using a "get out of jail free" card.

Another difficulty is the rule that allows repeated throws when doubles are thrown. Because of this rule, it is possible to pass through three positions on the board in a single turn. Hence the authors took a *throw* of the dice as a basic unit of time rather than a player's turn. In view of this, the states of the Markov chain were described by an ordered pair (i,j), with i corresponding to the position on the board and $j = 1$, 2, or 3 depending on which throw of the turn is coming up. For example, the state $(5,1)$ corresponds to being on the Reading Railroad with the first throw of a turn coming up. The jail position (or "in jail" position to which one is sent by the "go to jail" position) is somewhat different. However, under the RIJ rule, there are still three states associated with the jail position corresponding to the number of turns for which the jail stay has lasted. Hence under the RIJ rule, there are 120 possible states, and there are 118 possible states under the LJ rule.

Finally there is a problem with the Community Chest and Chance cards. If the cards are placed at the bottom of the deck after each use, then the process is not Markovian because of the dependence on past events. To get around this problem, the rules could be modified to require the entire deck be shuffled rather than returning the drawn card to the bottom of the deck. The "get out of jail free" cards can also be replaced immediately with a system of "credit" being used rather than the actual cards themselves.

If all of the above assumptions are made and if the game is played under the RIJ rule, then the probabilities for the 120×120 transition matrix can be calculated mathematically. For example, if the states are written as $(1,1)$, $(2,1)$,...$(39,1)$, $(1,2)$, $(2,2)$,...$(39,2)$, $(1,3)$, $(2,3)$,...$(39,3)$, $(J,1)$, $(J,2)$, and $(J,3)$, then $p_{1,5} = P[X_n = (5,1)|X_{n-1} = (1,1)] = 2/36$ and $p_{1,44} = P[X_n = (5,2)|X_{n-1} = (1,1)] = 1/36$. Some transition probabilities are more difficult to calculate because of the effect of Chance and Community Chest cards. The entire matrix can be written in the form

$$P = \begin{bmatrix} A & B & 0 & J \\ A & 0 & B & J \\ C & 0 & 0 & J \\ K & K & 0 & L \end{bmatrix}$$

where each of the nine matrices in the upper left are 39×39 and the matrices J, K, and L are 39×3, 3×39, and 3×3, respectively.

The authors calculated the long run probabilities of being in various positions under the RIJ and the LJ rule. As one might guess, the probability of being in positions just after the jail position is somewhat higher than the average and the probability of being in positions just after the "go to jail" position are somewhat lower than the average (about 20% difference in both cases). (A complete table of these probabilities can be found in the original paper.)

The expected income from the bank per turn was calculated under the RIJ and LJ rules and is given in Table III.6.4.

Table III.6.4.   Expected Income from Bank (Dollars per Turn).

	RIJ	LJ
Salary (Pass Go)	38.48	34.64
Income and Luxury Taxes	−7.06	−7.49
Community Chest	1.94	2.03
Chance	1.25	1.29
Get out of jail charge	−1.29	−2.35
Total	33.37	34.64

Finally, the expected property group income, with hotels was calculated under the RIJ and LJ rules and is given in Table III.6.5.

Table III.6.5.   Expected Property-group Income, with hotels (dollars per opponents' turn)

Property Group	RIJ	LJ
Mediterranean-Baltic	16.8	17.8
Oriental-Vermont-Connecticut	42.3	44.9
St. Charles-States-Virginia	68.1	71.5
St. James-Tennessee-New York	95.4	101.2
Kentucky-Indiana-Illinois	103.6	110.9
Atlantic-Ventnor-Marvin Gardens	103.7	110.3
Pacific-N. Carolina-Pennsylvania	114.8	121.9
Park Place-Boardwalk	95.8	101.4
Railroads (one owner)	27.3	29.2
Utilities (one owner)	4.4	4.5

If the reader would like to read further on the strategies of playing Monopoly, he is referred to the article "How to Win at Monopoly" by Irwin Hentzel, *Saturday Review of the Sciences*, pp. 44–48, April, 1973.

Application III.6.5.   "A Finite Model of Mobility," by Neil Henry, Robert McGinnis, and Heinrich Tegtmeyer, *Journal of Mathematical Sociology*, **1**, pp. 107–118, 1971.

Stationary Markov chains have been used by sociologists as a model for studying human mobility. They have been used in attempts to model short run moves among industrial categories, intergenerational occupational mobility, and human migration. In studying human migration, sociologists have used various geographical locations as the states and estimated the transition probabilities using past data. Unfortunately, in this case the Markov chain has not proven to be a good model of mobility.

One serious problem with the Markov chain model is the assumption that all persons in a particular location have the same probability of moving to other locations regardless of their past histories. Actually, people who have lived in a particular location for a relatively long period of time develop strong ties (family, social, economic, etc.) to that location and are less likely to move than are people who have lived in that location for a relatively short period of time.

The Cornell Mobility Model (McGinnis, 1968) attempted to remedy this defect in the simple Markov chain model by accounting for the duration of residence in a given location. In the following, the assumptions and definitions used in the Cornell Mobility Model are given.

Let the set of locations be denoted by $S = \{s_i : i = 1, 2, \ldots, m\}$ and the duration of residence by $r = 1, 2, 3, \ldots$. Then the states of the chain can be labeled by the ordered pair (s_i, r). For example, the pair $(s_2, 3)$ would represent the state of being in location s_2 and being there for the third consecutive time unit. With this definition of states, it is clear that the transition must be from (s_i, r) at time n to either $(s_i, r+1)$ or to $(s_j, 1)$, $j \neq i$. The transition probabilities can be defined by

$$_r p_{ii} = P[X_n = (s_i, r+1) | X_{n-1} = (s_i, r)]$$

$$_r p_{ij} = P[X_n = (s_j, 1) | X_{n-1} = (s_i, r)], \qquad j \neq i.$$

Since the model is assumed to be stationary, these transition probabilities do not depend on n. This $m \times m$ matrix can be denoted by $_r P$.

An important assumption of the Cornell Mobility Model is called the Axiom of Cumulative Inertia. The axiom states that the tendency to remain in a given location increases as the duration of prior occupancy increases. Mathematically, the axiom requires that

$$_r p_{ii} \leqslant \ _{r+1} p_{ii} \leqslant 1.$$

The last assumption for this model is that the proportion of movers from s_i who go to s_j is independent of r, that is,

$$\frac{_r p_{ij}}{1 - \ _r p_{ii}} = \frac{_1 p_{ij}}{1 - \ _1 p_{ii}} \qquad \text{for all } i \neq j \text{ and all } r.$$

A problem with this model is that the state space, and hence the transition matrix, is infinite since r is allowed to range through the positive integers. As an attempt to solve this problem, the authors of this paper considered the additional assumption that the probability of leaving a location becomes constant after a specified period of continuous residence there. Mathematically this assumption implies the existence of an integer v such that $_r P = \ _v P$ for all $r > v$. This modification allows the migration to be treated by a finite Markov chain with states (s_i, r), $i = 1, \ldots m$ and $r = 1, 2, \ldots v$.

Using this modified model, two cases were considered. In the first case it was assumed that $_v P = I$, the identity matrix. This would imply that persons with a duration status of $r \geqslant v$ would never leave their location of residence. In the second case it was assumed that $_v P \neq I$ (in fact it was assumed that $_v P$ has no ones on the diagonal), so that there would always be a chance of moving to a new location. It should be noted that the first alternative leads to an absorbing Markov chain, while the second leads to an irreducible aperiodic (ergodic) Markov chain.

If $_rD$ is defined to be the diagonal part of $_rP$, that is,

$$_rd_{ii} = {_rp_{ii}}$$

$$_rd_{ij} = 0, \qquad i \neq j$$

and if $_rM = {_rP} - {_rD}$ is the off-diagonal part of $_rP$, then the $mv \times mv$ transition matrix for the finite Markov chain using the modified model is the matrix given in Table III.6.6.

Table III.6.6. Transition matrix using modified model.

Duration →		1	2	3	...	v
↓	State	$(s_1,1)...(s_m,1)$	$(s_1,2)...(s_m,2)$	$(s_1,3)...(s_m,3)$	...	$(s_1,v)...(s_m,v)$
1	$(s_1,1)$ ⋮ $(s_m,1)$	$_1M$	$_1D$	0	...	0
2	$(s_1,2)$ ⋮ $(s_m,2)$	$_2M$	0	$_2D$	...	0
	⋮ $(s_1,v-1)$	⋮	⋮	⋮	⋮	
$v-1$	⋮ $(s_m,v-1)$	$_{v-1}M$	0	0	...	$_{v-1}D$
v	(s_1,v) ⋮ (s_m,v)	$_vM$	0	0	...	$_vD$

In the case where $_vP = I$, it follows that $_vD = I$ and $_vM = 0$. It is easy to see that the absorbing chain defined by this case has the property that it is absorbing with respect to duration, but no single location is absorbing. After a long time, an individual will "come to rest," but this may occur in any location. For this case absorption probabilities and the expected number of moves a person makes in a lifetime are calculated using the fundamental matrix, $(I - Q)^{-1}$.

In the case where $_vP \neq I$, that is the case where the chain is ergodic, the long run distribution, π, is calculated.

As stated earlier, the transition probabilities for this model were estimated from past data. Two independent studies indicate that selected migration data conform reasonably well to these models. However, the authors did point out some limitations of this model. In particular, the model does not account for deaths or births into the population, which is a serious limitation in certain cases. There are also individual, locational, and system characteristics that might be included in the model such as age, changes in marital status or family size, occupational opportunities, cost of living, and distances between locations. The authors of this paper felt that more work is needed on this problem. In particular they felt that the Markov chain model needs to be further modified to allow for the inclusion of these new variables without reaching a level of complexity that limits its practical application in social research.

Application III.6.6. "Reaching a Consensus," by M. H. DeGroot, *Journal of the American Statistical Association*, **69**, pp. 118–121, 1974.

In this paper DeGroot did not use a Markov chain, but the method he proposed for reaching a consensus does lead to a stochastic matrix. By analyzing the stochastic matrix it is possible to determine whether or not a consensus can be reached, and how to reach a consensus if one in fact can be reached.

Consider the problem of having k individuals who must act as a group or committee in order to give a subjective probability distribution for a parameter θ. We assume that each person has his own subjective probability distribution for θ, say F_i for the ith person. The problem is to "pool" the ideas of all of the individuals and arrive at a single distribution with which all individuals are "satisfied." [One simple way of doing this might be to take the "average" distribution by using $A = (F_1 + F_2 + \cdots + F_k)/k$.]

The author suggested the following approach to the problem. Consider one particular individual in the group, say the ith individual, and assume he knows or is made aware of the background, ability, expertise, and so on of the other $k - 1$ individuals in the group. Given this awareness, assume he is willing to revise his subjective distribution by considering the opinions of the remaining group members according to the following scheme: Let $p_{ij}, j = 1, \ldots, k$ be the "weight" assigned by the ith individual to the distribution proposed by the jth individual, subject to $p_{ij} \geq 0$ and $\sum_{i=1}^{k} p_{ij} = 1$. (These weights would reflect the relative confidence in the opinions of the individuals in the group. An individual with a large ego might set $p_{ii} = 1$, while another person feeling that all were equally qualified might set $p_{ij} = 1/k$ for all j.) The revised distribution would be given by

$$F_i^{(1)} = \sum_{j=1}^{k} p_{ij} F_j.$$

If all of the members of the group followed the same procedure, then the row vector of revised probability distributions, denoted by $F^{(1)} = (F_1^{(1)}, F_2^{(1)}, \ldots, F_k^{(1)})$ could be found from the row vector F of original probability distributions, $F = (F_1, F_2, \ldots, F_k)$, and the stochastic matrix P

$= (p_{ij})$ by the equation

$$F^{(1)\prime} = PF'.$$

If it is assumed that each individual retains the same set of weights p_{ij}, then a second revision would yield

$$F^{(2)\prime} = PF^{(1)\prime} = P^2 F'$$

and in general $F^{(n)\prime} = P^n F'$.

One can see from this last equation that a consensus will be reached if and only if $\lim_{n \to \infty} P^n$ exists, or in other words if and only if P is ergodic. If P is in fact ergodic and if $\pi = (\pi_1, \pi_2, \ldots, \pi_k)$ is the stationary distribution for P, then the consensus probability distribution for θ is given by

$$F^* = \sum_{i=1}^{k} \pi_i F_i.$$

CHAPTER IV

The Algebraic Approach to Markov Chains

In previous chapters we have seen that the behavior of a stationary Markov chain is determined by a starting vector and a transition matrix P. More precisely, in Chapter III we saw that under certain conditions, each row of P^n converges to a vector π which is a left eigenvector of P corresponding to the eigenvalue 1. The elements of the eigenvector π, say π_i, give the probability that the chain will be in state i "after a long time."

In this chapter we will see that the consideration of all of the eigenvalues of a stochastic matrix P can give us a good deal of information about the nature of a Markov chain having P as its transition matrix. In particular we will see that the composition of the set of all eigenvalues of P is directly related to the periodicity and the number of persistent classes associated with P (Theorems IV.2.4 and IV.2.5). Under certain simplifying assumptions we will see how the eigenvalues and corresponding eigenvectors can be used to give a simple expression for P^k. This expression for P^k also will allow us to see why P^k converges to a matrix with constant rows all equal to a certain eigenvector.

Since the proofs of the theorems in this chapter require a greater knowledge of linear algebra than has been needed in this book up to this point, we begin by stating some basic definitions and simply state without proof those pertinent theorems whose proofs are somewhat involved. We illustrate the definitions and theorems with examples. It is our intention to illustrate what can be done using the algebraic approach to Markov chains and to relate this approach to the classical approach given in Chapter III. The interested reader can find a more detailed discussion of the topics in this chapter in Karlin (1968) or Feller (1968).

SECTION 1 **ALGEBRAIC REVIEW**

Let A be an $n \times n$ matrix whose elements are real numbers and let $\mathbf{x}$ and $\mathbf{y}$ be row vectors.

Definition IV.1.1. *A number λ is said to be an* eigenvalue *of a matrix A if there exists some nonzero vector $\mathbf{x}$ satisfying $\mathbf{x}A = \lambda\mathbf{x}$. Such a vector $\mathbf{x}$ is called a* left eigenvector *of A corresponding to the eigenvalue λ. Similarly a nonzero column vector $\mathbf{y}'$ satisfying $A\mathbf{y}' = \lambda\mathbf{y}'$ is called a* right eigenvector *of A.*

Example IV.1.1. The vector $\mathbf{y}' = (3, -1, 3)'$ is a right eigenvector of

$$A = \begin{bmatrix} 5 & -6 & -6 \\ -1 & 4 & 2 \\ 3 & -6 & -4 \end{bmatrix}$$

corresponding to the eigenvalue 1. The vector $\mathbf{x}_1 = (-1, 2, 2)$ is a left eigenvector corresponding to the eigenvalue 1.

It is easy to see that any constant multiple of $\mathbf{x}_1$ is also a left eigenvector of A corresponding to the eigenvalue 1. We might ask if there are any left eigenvectors corresponding to $\lambda = 1$ that are not multiples of $\mathbf{x}_1$. In this case, there are not. However, the reader can easily verify that $\mathbf{x}_2 = (1, 6, 1)$ and $\mathbf{x}_3 = (1, 0, -1)$ are two left eigenvectors corresponding to $\lambda = 2$ that are not multiples of each other. Since they are not multiples of each other it follows that no nontrivial linear combination of $\mathbf{x}_2$ and $\mathbf{x}_3$ can be equal to the zero vector. This notion can be generalized to more than two vectors and so we can give the following definitions.

Definition IV.1.2. *A set of vectors $\mathbf{x}_1, \mathbf{x}_2, \ldots, \mathbf{x}_k$ is said to be* linearly independent *if $\alpha_1\mathbf{x}_1 + \alpha_2\mathbf{x}_2 + \cdots + \alpha_k\mathbf{x}_k = 0$ implies $\alpha_1 = \alpha_2 = \cdots = \alpha_k = 0$.*

Definition IV.1.3. *The* multiplicity *of an eigenvalue λ is equal to the number of independent left eigenvectors corresponding to λ.*

Since in Example IV.1.1 the vectors $\mathbf{x}_2$ and $\mathbf{x}_3$ are independent, the multiplicity of $\lambda = 2$ is two. Similarly, it follows that the multiplicity of $\lambda = 1$ is one.

It is shown in most textbooks on linear algebra that the eigenvalues of a matrix A can be found by solving $|A - \lambda I| = 0$ (where $|M|$ denotes the determinant of M). Evaluation of the determinant of $(A - \lambda I)$ leads to an n degree polynomial in λ called the characteristic polynomial of A. The roots of this polynomial will be the eigenvalues of A.

If A is as defined in Example IV.1.1, then the reader can show that the characteristic polynomial is

$$\lambda^3 - 5\lambda^2 + 8\lambda - 4 = (\lambda - 1)(\lambda - 2)^2.$$

If we consider the set of all eigenvalues of a matrix, we may find that there is an eigenvalue, λ_0 say, which is of multiplicity one and which is larger in absolute value than any other eigenvalue. Such an eigenvalue is called the *dominant eigenvalue* or the *dominant root*. Of course, not every matrix has a dominant eigenvalue, as can be seen by considering the matrix of Example IV.1.1.

Most readers are no doubt familiar with the fact that a multiplicative identity for matrices exists and that some, but not all, $n \times n$ matrices have a multiplicative inverse. A necessary and sufficient condition for the inverse of a matrix A to exist is that the rows of A be linearly independent. This fact will be used in proving Theorem IV.2.6.

EXERCISES

1. Find the eigenvalues and their multiplicities for the following matrices. Determine the dominant root if it exists.

(a) $A = \begin{pmatrix} 1 & 0 & 0 \\ \frac{1}{2} & 0 & \frac{1}{2} \\ \frac{1}{3} & \frac{1}{3} & \frac{1}{3} \end{pmatrix}$　　(d) $D = \begin{pmatrix} \frac{1}{3} & \frac{1}{3} & \frac{1}{3} \\ \frac{1}{2} & \frac{1}{2} & 0 \\ 0 & 1 & 0 \end{pmatrix}$

(b) $B = \begin{pmatrix} 0 & \frac{1}{2} & 0 & \frac{1}{2} \\ \frac{1}{3} & 0 & \frac{2}{3} & 0 \\ 0 & 0 & 0 & 1 \\ 1 & 0 & 0 & 0 \end{pmatrix}$　　(e) $E = \begin{pmatrix} 0 & 1 & 0 \\ 0 & 0 & 1 \\ 1 & 0 & 0 \end{pmatrix}$

(c) $C = \begin{pmatrix} \frac{1}{2} & 0 & \frac{1}{2} \\ 0 & 1 & 0 \\ \frac{1}{4} & 0 & \frac{3}{4} \end{pmatrix}$

2. Find the inverse of the matrix defined in Example IV.1.1.
3. Show that the characteristic polynomial of the matrix defined in Example IV.1.1 is $\lambda^3 - 5\lambda^2 + 8\lambda - 4$.

SECTION 2 **THEOREMS FOR CLASSIFICATION OF FINITE MARKOV CHAINS**

As in Section 1, we restrict our attention in this section to finite matrices but in this section we will further restrict our attention to stochastic matrices. We begin with the following theorem.

Theorem IV.2.1.

All finite stochastic matrices P have 1 as an eigenvalue. There exist non-negative eigenvectors corresponding to $\lambda = 1$.

Proof. Since the elements of each row of P must sum to 1, it is easy to see that $\mathbf{y}' = (1, 1, \ldots, 1)'$ satisfies $P\mathbf{y}' = 1 \cdot \mathbf{y}'$. Hence by Definition IV.1.1 the number 1 is an eigenvalue for P. Of course $\mathbf{y}'$ is a right eigenvector of P. It is somewhat more interesting to try to find a left eigenvector corresponding to $\lambda = 1$. In fact we will try to find a left eigenvector with non-negative components that sum to one.

We know (Lemma III.2.1) that all finite chains must have at least one positive persistent state. This state generates a closed irreducible subclass of S and the chain restricted to this class is an irreducible positive persistent chain. For such a chain we know there exists an invariant probability vector. Assume P is $n \times n$ and rewrite P into block form with all the states of the above-mentioned closed subclass placed in the upper left hand corner of P. P then takes the form

$$P^* = \begin{pmatrix} P_1 & 0 \\ R & Q \end{pmatrix}$$

where P_1 is the probability transition matrix corresponding to a closed, irreducible, positive persistent subclass. (Note that P^* and P have the same eigenvalues—Exercise 5.) Let $\boldsymbol{\pi} = (\pi_1, \pi_2, \ldots, \pi_k)$ be the invariant probability vector for P_1. That is, $\boldsymbol{\pi} P_1 = \boldsymbol{\pi}$. Define $\boldsymbol{\psi} = (\pi_1, \pi_2, \ldots, \pi_k, 0, 0, \ldots, 0)$ and note that $\boldsymbol{\psi} P = \boldsymbol{\psi}$. Hence $\boldsymbol{\psi}$ is a left eigenvector for P corresponding to $\lambda = 1$ and $\sum_{i=1}^n \psi_i = 1$. ▲

Not only is 1 an eigenvalue of the finite stochastic matrix P, but it is in fact the largest eigenvalue of P. We will not prove this fact here but rather refer to the Perron–Frobenius theorems which apply to the class of matrices with non-negative entries. The results are usually given as two separate theorems; one for the case where A^m has non-negative entries for all m and one for the case where A^m has all entries strictly positive for some m. For formal statements and proofs of these theorems, the reader is referred to Gantmacher (1959), Karlin (1968), or Seneta (1973).

Theorem IV.2.2 (Perron–Frobenius).

Let A be an $n \times n$ matrix with non-negative entries. Then

(i) A has a positive eigenvalue λ_0 with corresponding left eigenvector $\mathbf{x}_0$ such that $\mathbf{x}_0$ is non-negative and nonzero

(ii)   if λ is any other eigenvalue of A, $|\lambda| \leqslant \lambda_0$
(iii) if λ is an eigenvalue of A and $|\lambda| = \lambda_0$, then $\eta = \lambda/\lambda_0$ is a root of unity and $\eta^k \lambda_0$ is an eigenvalue of A for $k = 0, 1, 2 \ldots$.

Theorem IV.2.3   (Perron–Frobenius).

Let A be an $n \times n$ matrix with non-negative entries such that A^m has all positive entries for some m. Then

(i) A has a positive eigenvalue λ_0 with corresponding left eigenvector $\mathbf{x}_0$ where the entries of $\mathbf{x}_0$ are positive
(ii)   if λ is any other eigenvalue of A, $|\lambda| < \lambda_0$
(iii) λ_0 has multiplicity one.

We now use these two theorems in order to show that $\lambda = 1$ is the largest eigenvalue for P. Since P is a stochastic matrix it is true that all of its entries are non-negative, so that the conclusion of Theorem IV.2.2 will hold. It also follows from Theorem IV.2.1 that $\lambda = 1$ is an eigenvalue of P and the corresponding left eigenvector is non-negative, and so $\lambda = 1$ is a candidate for the λ_0 of Theorem IV.2.2. We will show that no other eigenvalue of P has a nonzero left eigenvector that is non-negative, so it must be that $\lambda_0 = \lambda = 1$. To see this, assume that $\mathbf{x}$ is a non-negative, nonzero left eigenvector of P corresponding to λ. This means $\lambda \mathbf{x} = \mathbf{x} P$, which implies that

$$\lambda x_i = \sum_{j=1}^{n} x_j p_{ji}, \qquad i = 1, 2, \ldots, n. \tag{2.1}$$

Since P is stochastic we know $\sum_{i=1}^{n} p_{ji} = 1$, so summing both sides of (2.1) we get

$$\lambda \sum_{i=1}^{n} x_i = \sum_{i=1}^{n} \sum_{j=1}^{n} x_j p_{ji} = \sum_{j=1}^{n} \sum_{i=1}^{n} x_j p_{ji} = \sum_{j=1}^{n} x_j.$$

Since the vector $\mathbf{x}$ was taken to be non-negative and nonzero, it must be that $\lambda = 1$. Furthermore we can see that for all eigenvalues of P other than $\lambda = 1$, it must be that the corresponding left eigenvectors must have components that sum to zero. Hence it follows from (ii) of Theorem IV.2.2 that all eigenvalues λ of P are such that $|\lambda| \leqslant \lambda_0 = 1$, that is, $\lambda_0 = 1$ is the largest eigenvalue for P.

If P is the transition matrix corresponding to an irreducible aperiodic Markov chain, then P^m has all positive entries for some m (Exercise 3).

Hence it follows from Theorem IV.2.3 that for irreducible aperiodic stochastic matrices the eigenvalue $\lambda = 1$ is dominant.

The next two theorems show how the eigenvalue 1 of a stochastic matrix is related to irreducibility and periodicity of the Markov chain determined by that matrix.

Theorem IV.2.4.

If P is the transition matrix for a finite Markov chain, then the multiplicity of the eigenvalue 1 is equal to the number of irreducible closed subsets of the chain.

Proof. From Chapter III we know that P can be rewritten into block form with each block, P_i, $i = 1, 2, \ldots, m$, corresponding to a closed irreducible subset of S and the last block, Q, corresponding to the transient states.

That is,

$$
P = \begin{bmatrix}
P_1 & 0 & \cdots & 0 & 0 \\
0 & P_2 & \cdots & 0 & 0 \\
\vdots & \vdots & & \vdots & \vdots \\
0 & 0 & \cdots & P_m & 0 \\
R_1 & R_2 & \cdots & R_m & Q
\end{bmatrix}.
$$

Each of these blocks, P_i, corresponds to a state space, C_i, for an irreducible, positive persistent chain. Hence by results in Chapter III there exists a left eigenvector, $\mathbf{x}_i = (0 \ldots 0, x_1^{(i)}, x_2^{(i)}, \ldots, x_{n_i}^{(i)}, 0, \ldots, 0)$ with positive entries in the coordinates corresponding to C_i and zeros elsewhere with $\mathbf{x}_i P = \mathbf{x}_i$ for $i = 1, 2, \ldots, m$. Since the C_i's are disjoint by construction, the $\mathbf{x}_i$'s have positive entries in nonoverlapping coordinates, so they are linearly independent. Hence the multiplicity of the eigenvalue 1 is at least equal to the number of closed irreducible subsets of S.

In order to show that the multiplicity is at most m, we take $\mathbf{y}$ to be an arbitrary left eigenvector of P corresponding to the eigenvalue 1 and show $\mathbf{y}$ can be written as a linear combination of the $\mathbf{x}_i$'s, $i = 1, 2, \ldots, m$. If $\mathbf{y}P = \mathbf{y}$, then $\mathbf{y}P^k = \mathbf{y}$ for all $k = 1, 2, \ldots$ so $\sum_{i=1}^{n} y_i p_{ij}^{(k)} = y_j$ for all k. If j is a transient state, then $\lim_{k \to \infty} p_{ij}^{(k)} = 0$, so all the coordinates of $\mathbf{y}$ corresponding to transient states are zero. The equation $\sum_{i=1}^{n} y_i p_{ij} = y_j$ can now be rewritten summing over only the persistent states, so

$$
\sum_{k=1}^{m} \sum_{i \in C_k} y_i p_{ij} = y_j \qquad \text{for} \qquad j \in \bigcup_{k=1}^{m} C_k.
$$

Now $p_{ij} = 0$ if i and j belong to distinct closed subsets so $\Sigma_{i \in C_k} y_i p_{ij} = y_j$ for $j \in C_k$, $k = 1, 2, \ldots, m$. But for an irreducible positive persistent chain the equation $\Sigma_{i \in C_k} y_i p_{ij} = y_j$ has a unique solution up to scalar multiples (Theorem III.2.2). Hence we get $\mathbf{y} = \Sigma_{k=1}^{m} a_k \mathbf{x}_k$. Therefore the multiplicity of the eigenvalue one is equal to the number of closed irreducible subsets of S. ▲

Theorem IV.2.5.

If P is the transition matrix of an irreducible periodic Markov chain with period d, then the dth roots of unity are eigenvalues of P. Further, each of these eigenvalues is of multiplicity one and there are no other eigenvalues of modulus one.

The proof of this theorem will not be given here. The interested reader is referred to Karlin (1968) for a rather detailed proof.

Corollary IV.2.1. If P is the transition matrix of a Markov chain, then any eigenvalue of P of modulus 1 is a root of unity. The dth roots of unity are eigenvalues of P if and only if P has a persistent subclass of period d. The multiplicity of each collection of dth roots of unity is the number of persistent subclasses of period d.

Proof. Since $\lambda \mathbf{x} = \mathbf{x} P$ implies $\lambda^m \mathbf{x} = \mathbf{x} P^m$ or $\lambda^m x_j = \Sigma_{i=1}^{n} x_i p_{ij}^{(m)}$ we have $x_j = 0$ for all transient states j since $\lim_{m \to \infty} p_{ij}^{(m)} = 0$. Hence we may restrict our attention to the closed irreducible subsets of S and apply the previous theorem to each of these separately. ▲

To conclude this section we give a means whereby, under certain conditions, it is relatively easy to calculate powers of P. This technique consists of using the spectral representation or the spectral decomposition for a matrix. The linear algebra text by de Pillis (1969) has a good discussion of this technique. Rao (1965) is also a good reference for those readers desiring a bit of a statistical slant. We will begin by giving a theorem that holds for general finite matrices and then will apply the results to stochastic matrices.

Theorem IV.2.6.

Let A be an $n \times n$ matrix having n linearly independent left eigenvectors $\mathbf{x}_1, \mathbf{x}_2, \ldots, \mathbf{x}_n$ corresponding to eigenvalues $\lambda_1, \lambda_2, \ldots, \lambda_n$. Define matrices L

and Λ by

$$L = \begin{bmatrix} \mathbf{x}_1 \\ \mathbf{x}_2 \\ \vdots \\ \mathbf{x}_n \end{bmatrix} \qquad \Lambda = \begin{bmatrix} \lambda_1 & 0 & 0 & \ldots & 0 \\ 0 & \lambda_2 & 0 & \ldots & 0 \\ \vdots & \vdots & \vdots & & \vdots \\ 0 & 0 & 0 & \ldots & \lambda_n \end{bmatrix}.$$

Then $A = L^{-1}\Lambda L$.

Proof. By our earlier remarks we know that if the rows of L, which are the left eigenvectors of A, are independent, then L^{-1} exists. Since we have

$$LA = \begin{bmatrix} \mathbf{x}_1 A \\ \mathbf{x}_2 A \\ \vdots \\ \mathbf{x}_n A \end{bmatrix} = \begin{bmatrix} \lambda_1 \mathbf{x}_1 \\ \lambda_2 \mathbf{x}_2 \\ \vdots \\ \lambda_n \mathbf{x}_n \end{bmatrix} = \Lambda L$$

and since L is invertible, we get that

$$L^{-1}LA = A = L^{-1}\Lambda L. \qquad \blacktriangle$$

This representation makes it easy to calculate powers of A since

$$A^2 = L^{-1}\Lambda L \ L^{-1}\Lambda L = L^{-1}\Lambda^2 L$$

and in general $A^k = L^{-1}\Lambda^k L$. Of course since Λ is a diagonal matrix we have

$$\Lambda^k = \begin{bmatrix} \lambda_1^k & 0 & 0 & \ldots & 0 \\ 0 & \lambda_2^k & 0 & \ldots & 0 \\ \vdots & \vdots & \vdots & & \vdots \\ 0 & 0 & 0 & \ldots & \lambda_n^k \end{bmatrix}.$$

An alternative representation for powers of A can be given in a slightly different way. In Exercise 2 the reader is asked to show that the columns of L^{-1} are right eigenvectors of the matrix A. If we denote the columns of

L^{-1} by $y_1', y_2', \ldots, y_n'$, then we get that

$$A = L^{-1}\Lambda L = (y_1', y_2', \ldots, y_n')\Lambda \begin{bmatrix} x_1 \\ x_2 \\ \vdots \\ x_n \end{bmatrix} = \sum_{i=1}^{n} \lambda_i y_i' x_i \qquad (2.2)$$

and

$$A^k = \sum_{i=1}^{n} \lambda_i^k y_i' x_i.$$

Example IV.2.1   Let

$$A = \begin{bmatrix} 4 & 0 & 0 \\ 0 & 2 & -1 \\ 0 & -1 & 2 \end{bmatrix}.$$

Then the characteristic polynomial of A is

$$|A - \lambda I| = -\lambda^3 + 8\lambda^2 - 19\lambda + 12 = (4 - \lambda)(\lambda - 3)(\lambda - 1)$$

so the eigenvalues of A are $\lambda_1 = 4$, $\lambda_2 = 3$, and $\lambda_3 = 1$. The reader can verify that the corresponding left eigenvectors can be taken to be $x_1 = (1, 0, 0)$, $x_2 = (0, 1, -1)$, and $x_3 = (0, 1, 1)$. Since these vectors are independent we have

$$L = \begin{bmatrix} 1 & 0 & 0 \\ 0 & 1 & -1 \\ 0 & 1 & 1 \end{bmatrix} \quad \text{and} \quad L^{-1} = \begin{bmatrix} 1 & 0 & 0 \\ 0 & \frac{1}{2} & \frac{1}{2} \\ 0 & -\frac{1}{2} & \frac{1}{2} \end{bmatrix}.$$

Direct multiplication shows that

$$L^{-1}\Lambda L = \begin{bmatrix} 1 & 0 & 0 \\ 0 & \frac{1}{2} & \frac{1}{2} \\ 0 & -\frac{1}{2} & \frac{1}{2} \end{bmatrix} \begin{bmatrix} 4 & 0 & 0 \\ 0 & 3 & 0 \\ 0 & 0 & 1 \end{bmatrix} \begin{bmatrix} 1 & 0 & 0 \\ 0 & 1 & -1 \\ 0 & 1 & 1 \end{bmatrix}$$

$$= \begin{bmatrix} 4 & 0 & 0 \\ 0 & \frac{3}{2} & \frac{1}{2} \\ 0 & -\frac{3}{2} & \frac{1}{2} \end{bmatrix} \begin{bmatrix} 1 & 0 & 0 \\ 0 & 1 & -1 \\ 0 & 1 & 1 \end{bmatrix} = \begin{bmatrix} 4 & 0 & 0 \\ 0 & 2 & -1 \\ 0 & -1 & 2 \end{bmatrix} = A.$$

The reader can verify that the columns of L^{-1}, namely $\mathbf{y}_1' = (1, 0, 0)'$, $\mathbf{y}_2' = (0, \frac{1}{2}, -\frac{1}{2})'$, and $\mathbf{y}_3' = (0, \frac{1}{2}, \frac{1}{2})'$, are right eigenvectors of A corresponding to eigenvalues 4, 3, and 1, respectively. Using the representation of (2.2) we see that

$$A = \sum_{i=1}^{3} \lambda_i \mathbf{y}_i' \mathbf{x}_i = 4 \begin{bmatrix} 1 & 0 & 0 \\ 0 & 0 & 0 \\ 0 & 0 & 0 \end{bmatrix} + 3 \begin{bmatrix} 0 & 0 & 0 \\ 0 & \frac{1}{2} & -\frac{1}{2} \\ 0 & -\frac{1}{2} & \frac{1}{2} \end{bmatrix} + \begin{bmatrix} 0 & 0 & 0 \\ 0 & \frac{1}{2} & \frac{1}{2} \\ 0 & \frac{1}{2} & \frac{1}{2} \end{bmatrix}.$$

It is easy to see how useful Theorem IV.2.6 can be for $n \times n$ stochastic matrices. If we know that P possesses a set of n linearly independent left eigenvectors then P^k can rather easily be determined from $L^{-1} \Lambda^k L$ or $\sum_{i=1}^{n} \lambda_i^k \mathbf{y}_i' \mathbf{x}_i$. Furthermore, if P is the $n \times n$ transition matrix for a Markov chain having a single aperiodic closed set, then $\lambda_1 = 1$ will be the dominant eigenvalue, so that the remaining eigenvalues will have modulus strictly less than 1. This implies that $\lambda_1^k = 1$ and $\lambda_i^k \to 0$ as $k \to \infty$ if $i \neq 1$. Hence

$$P^k = L^{-1} \Lambda^k L = \sum_{i=1}^{n} \lambda_i^k \mathbf{y}_i' \mathbf{x}_i \to \mathbf{y}_1' \mathbf{x}_1 \quad \text{as } k \to \infty. \tag{2.3}$$

Even more can be said since when P is stochastic we know that $\mathbf{y}_1' = (c, c, \ldots, c)'$ is a right eigenvector of P corresponding to $\lambda_1 = 1$. In addition, using the notation established earlier we can write the left eigenvector corresponding to $\lambda = 1$ as $\mathbf{x}_1 = a\boldsymbol{\pi} = (a\pi_1, a\pi_2, \ldots, a\pi_n)$. Since P^k has rows that sum to one, it follows that $c = 1/a$ so that

$$P^k \to \begin{bmatrix} 1 \\ 1 \\ \vdots \\ 1 \end{bmatrix} (\pi_1, \pi_2, \ldots, \pi_n) = \begin{bmatrix} \pi_1 & \pi_2 & \cdots & \pi_n \\ \pi_1 & \pi_2 & \cdots & \pi_n \\ \vdots & \vdots & & \vdots \\ \pi_1 & \pi_2 & \cdots & \pi_n \end{bmatrix}.$$

An obvious consequence of this last relationship is that if $\mathbf{x}_0$ is a starting vector then $\lim_{k \to \infty} \mathbf{x}_0 P^k = \boldsymbol{\pi}$.

That the limit of $\mathbf{x}_0 P^k$ is the eigenvector $\boldsymbol{\pi}$ certainly comes as no surprise. However our intention in this chapter has not been primarily to prove "new" results, but to give alternative proofs to results that we have already found to be true. Nevertheless, if we do find an interesting result as a consequence of using the algebraic approach we certainly will not ignore it. One such interesting result is the following. If the conditions of Theorem IV.2.6 hold and if the eigenvalues of P are $\{1, \lambda_2, \lambda_3, \ldots, \lambda_n\}$ with $\max_{2 \leqslant i \leqslant n} |\lambda_i| < 1$, then the Euclidean distance between $\mathbf{x}_0 P^k$ and $\boldsymbol{\pi}$ goes to zero at a geometric rate as $k \to \infty$. To see this, we consider (2.3) and note that

$$\mathbf{x}_0 P^k = \mathbf{x}_0 \sum_{i=1}^{n} \lambda_i^k \mathbf{y}_i' \mathbf{x}_i = \mathbf{x}_0 \begin{pmatrix} 1 \\ 1 \\ \vdots \\ 1 \end{pmatrix} \boldsymbol{\pi} + \sum_{i=2}^{n} \lambda_i^k \mathbf{x}_0 \mathbf{y}_i' \mathbf{x}_i$$

$$= \boldsymbol{\pi} + \sum_{i=2}^{n} \lambda_i^k \mathbf{x}_0 \mathbf{y}_i' \mathbf{x}_i.$$

Subtracting $\boldsymbol{\pi}$ from both sides and using the triangle inequality we get

$$\| \mathbf{x}_0 P^k - \boldsymbol{\pi} \| = \| \sum_{i=2}^{n} \lambda_i^k \mathbf{x}_0 \mathbf{y}_i' \mathbf{x}_i \|$$

$$\leqslant \max_{2 \leqslant i \leqslant n} |\lambda_i|^k \sum_{i=2}^{n} \| \mathbf{x}_0 \mathbf{y}_i' \mathbf{x}_i \|. \tag{2.4}$$

(The norm used here is the standard Euclidean norm on R^{n-1}. See Exercise 4c of Section I.4.) The factor $\max_{2 \leqslant i \leqslant n} |\lambda_i|^k$ goes to zero geometrically as $k \to \infty$ while the remaining factor is a constant depending on $\mathbf{x}_0$.

Unfortunately the hypotheses of Theorem IV.2.6 will not be satisfied for all stochastic matrices. The following example gives a 3×3 stochastic matrix for which only two independent eigenvectors can be found.

Example IV.2.2. Consider a Markov chain with transition matrix

$$P = \begin{pmatrix} \frac{1}{3} & \frac{1}{3} & \frac{1}{3} \\ \frac{1}{3} & \frac{1}{3} & \frac{1}{3} \\ \frac{1}{6} & \frac{1}{2} & \frac{1}{3} \end{pmatrix}.$$

The eigenvalues for P are $1, 0$, and 0. Two corresponding left eigenvectors for P are $\boldsymbol{\psi}_1 = (5, 7, 6)$ and $\boldsymbol{\psi}_2 = (1, -1, 0)$. However, there is no third left eigenvector that is linearly independent of these two, hence the hypotheses of Theorem IV.2.6 are not satisfied.

EXERCISES

1. Show that P and P' have the same eigenvalues.

2. Show that if A and L are as defined in Theorem IV.2.6, then the columns of L^{-1} are right eigenvectors of the matrix A.

3. Show that if P is a finite transition matrix for an irreducible aperiodic Markov chain, some power of P is positive.

4. Find the eigenvalues and corresponding left and right eigenvectors for

$$P = \begin{pmatrix} \frac{1}{2} & \frac{1}{6} & \frac{1}{3} \\ \frac{1}{2} & \frac{1}{2} & 0 \\ \frac{1}{2} & 0 & \frac{1}{2} \end{pmatrix}.$$

Find $\lim_{n \to \infty} x P^n = \pi$. How large must n be in order to guarantee $\|x P^n - \pi\| < .01$ independently of the starting vector x.

5. Show that if P^* is a block form representation for P, then P^* and P have the same eigenvalues.

6. Find a stochastic matrix with eigenvalues $1, 1, 1, -1, 0, 0, \eta, \eta^2$ where $\eta = \exp(2\pi i/3)$ is a cube root of unity.

7. Repeat Exercise 4 for the matrices

$$P_1 = \begin{pmatrix} \frac{1}{2} & \frac{1}{2} & 0 \\ \frac{1}{2} & \frac{1}{2} & 0 \\ \frac{1}{3} & 0 & \frac{2}{3} \end{pmatrix}, \qquad P_2 = \begin{pmatrix} 0 & 1 & 0 \\ 0 & 0 & 1 \\ \frac{1}{4} & \frac{3}{4} & 0 \end{pmatrix}.$$

8. Let P be an $n \times n$ stochastic matrix with n linearly independent left eigenvectors, $\{x_1, x_2, \ldots, x_n\}$. Let x_0 be a starting vector and let $x_1 = \pi$. Prove that in writing x_0 as a linear combination of $\{x_1, x_2, \ldots, x_n\}$ the coefficient of x_1 must be one.

SECTION 3	SOME EXAMPLES

The following examples illustrate the use of the theorems that were given in Section IV.2.

Example IV.3.1. Let X_n be a Markov chain with transition matrix

$$P = \begin{pmatrix} 0 & 0 & 1 & 0 & 0 & 0 \\ 0 & 0 & 0 & 1 & 0 & 0 \\ 0 & 0 & 0 & 0 & 1 & 0 \\ 0 & 1 & 0 & 0 & 0 & 0 \\ 0 & 0 & 0 & 0 & 0 & 1 \\ 1 & 0 & 0 & 0 & 0 & 0 \end{pmatrix}.$$

The theorems of this chapter can be used to either classify the states of the associated Markov chain or to find the eigenvalues for P.

First P can be rewritten into block form as

$$P^* = \begin{bmatrix} 0 & 1 & 0 & 0 & 0 & 0 \\ 0 & 0 & 1 & 0 & 0 & 0 \\ 0 & 0 & 0 & 1 & 0 & 0 \\ 1 & 0 & 0 & 0 & 0 & 0 \\ 0 & 0 & 0 & 0 & 0 & 1 \\ 0 & 0 & 0 & 0 & 1 & 0 \end{bmatrix} = \begin{pmatrix} P_1 & 0 \\ 0 & P_2 \end{pmatrix}.$$

This reduction of P to block form is helpful whether one is interested in finding eigenvalues or classifying states. Note that P^* and P have the same eigenvalues (Exercise 5 of Section IV.2).

From the block form it is easy to see that there are two closed sets, the first consisting of 4 states and being of period 4, the second consisting of 2 states and being of period 2. Hence the eigenvalues of P are the fourth roots of unity and the square roots of unity.

On the other hand the eigenvalues of P^* might be calculated directly and used to classify the states. From the theory of determinants we know that if A can be partitioned as

$$A = \begin{pmatrix} B & 0 \\ 0 & C \end{pmatrix}$$

where B is $k \times k$ and C is $n-k \times n-k$, then $|A| = |B| \cdot |C|$. Using this result, the eigenvalues of P^* can be found by finding the eigenvalues of P_1 and P_2 separately.

Solving the determinantal equation we find (expanding by the first column)

$$|P_1 - \lambda I| = \begin{vmatrix} -\lambda & 1 & 0 & 0 \\ 0 & -\lambda & 1 & 0 \\ 0 & 0 & -\lambda & 1 \\ 1 & 0 & 0 & -\lambda \end{vmatrix} = -\lambda[-\lambda^3] - 1[1] = \lambda^4 - 1.$$

Similarly,

$$|P_2 - \lambda I| = \begin{vmatrix} -\lambda & 1 \\ 1 & -\lambda \end{vmatrix} = \lambda^2 - 1.$$

Therefore $|P - \lambda I| = (\lambda^4 - 1)(\lambda^2 - 1) = 0$ is solved by roots of $(\lambda^4 - 1) = 0$ and of $(\lambda^2 - 1) = 0$, namely 1, -1, i, $-i$ and 1, -1.

Considering all of the eigenvalues of P, we see that $\lambda = 1$ has multiplicity two; hence from Theorem IV.2.4 we can conclude that there are two irreducible closed sets. Considering the eigenvalues of P_1, the fourth roots of unity, we see that the first closed irreducible subset is of period 4 and similarly the second closed irreducible subset is of period 2.

Example IV.3.2. Let X_n be a Markov chain with transition matrix

$$P = \begin{bmatrix} \frac{1}{2} & \frac{1}{2} & 0 & 0 & 0 & 0 & 0 \\ \frac{1}{3} & \frac{1}{3} & \frac{1}{3} & 0 & 0 & 0 & 0 \\ 0 & \frac{1}{4} & \frac{3}{4} & 0 & 0 & 0 & 0 \\ 0 & 0 & 0 & \frac{1}{2} & \frac{1}{2} & 0 & 0 \\ 0 & 0 & 0 & 1 & 0 & 0 & 0 \\ 0 & 0 & \frac{1}{3} & 0 & 0 & \frac{1}{3} & \frac{1}{3} \\ 0 & 0 & 0 & \frac{1}{2} & 0 & \frac{1}{2} & 0 \end{bmatrix}.$$

Find the eigenvalues for P. In this case P is already in the block form

$$P = \begin{bmatrix} P_1 & 0 & 0 \\ 0 & P_2 & 0 \\ R_1 & R_2 & Q \end{bmatrix}.$$

In order to find the eigenvalues of P, we find the eigenvalues of P_1, P_2, and Q separately. The eigenvalues of P_1 are 1, $(7+\sqrt{73})/24$ and $(7-\sqrt{73})/24$. The eigenvalues of P_2 are 1 and $-\frac{1}{2}$. The eigenvalues of Q are $(1+\sqrt{7})/6$ and $(1-\sqrt{7})/6$. Hence we have the seven eigenvalues and from these we see that there must be two irreducible aperiodic closed sets. (Note that in general the techniques of Chapter III yield the faster way of classifying a Markov chain.)

The last example illustrates the use of Theorem IV.2.6.

Example IV.3.3. Let $\{X_n\}$ be a Markov chain with transition matrix

$$P = \begin{bmatrix} 1/2 & 7/24 & 1/6 & 1/24 \\ 1/6 & 11/24 & 1/6 & 5/24 \\ 1/2 & 7/24 & 1/6 & 1/24 \\ 1/6 & 5/24 & 1/6 & 11/24 \end{bmatrix}.$$

The reader can verify that the eigenvalues of P are 1, $\frac{1}{3}$, $\frac{1}{4}$, and 0. The corresponding left and right eigenvectors are

$$\pi = x_1 = \left(\tfrac{1}{3}, \tfrac{1}{3}, \tfrac{1}{6}, \tfrac{1}{6}\right), \qquad y_1 = (1,1,1,1)'$$

$$x_2 = \left(\tfrac{1}{2}, \tfrac{1}{2}, 0, -1\right), \qquad y_2 = (1, -1, 1, -1)'$$

$$x_3 = \left(0, \tfrac{1}{2}, 0, -\tfrac{1}{2}\right), \qquad y_3 = \left(-\tfrac{5}{3}, \tfrac{7}{3}, -\tfrac{5}{3}, \tfrac{1}{3}\right)'$$

$$x_4 = \left(\tfrac{1}{2}, 0, -\tfrac{1}{2}, 0\right), \qquad y_4 = \left(\tfrac{1}{3}, \tfrac{1}{3}, -\tfrac{5}{3}, \tfrac{1}{3}\right)'.$$

Using $P^k = \sum_{i=1}^{4} \lambda_i^k y_i' x_i$, we get that

$$
P^k =
\begin{bmatrix}
\tfrac{1}{3} & \tfrac{1}{3} & \tfrac{1}{6} & \tfrac{1}{6} \\
\tfrac{1}{3} & \tfrac{1}{3} & \tfrac{1}{6} & \tfrac{1}{6} \\
\tfrac{1}{3} & \tfrac{1}{3} & \tfrac{1}{6} & \tfrac{1}{6} \\
\tfrac{1}{3} & \tfrac{1}{3} & \tfrac{1}{6} & \tfrac{1}{6}
\end{bmatrix}
+ \left(\tfrac{1}{3}\right)^k
\begin{bmatrix}
\tfrac{1}{2} & \tfrac{1}{2} & 0 & -1 \\
-\tfrac{1}{2} & -\tfrac{1}{2} & 0 & 1 \\
\tfrac{1}{2} & \tfrac{1}{2} & 0 & -1 \\
-\tfrac{1}{2} & -\tfrac{1}{2} & 0 & 1
\end{bmatrix}
$$

$$
+ \left(\tfrac{1}{4}\right)^k
\begin{bmatrix}
0 & -\tfrac{5}{6} & 0 & \tfrac{5}{6} \\
0 & \tfrac{7}{6} & 0 & -\tfrac{7}{6} \\
0 & -\tfrac{5}{6} & 0 & \tfrac{5}{6} \\
0 & \tfrac{1}{6} & 0 & -\tfrac{1}{6}
\end{bmatrix}.
$$

If we carry calculations to four decimal places, we get for $k = 5$

$$
P^5 =
\begin{bmatrix}
.3354 & .3346 & .1667 & .1634 \\
.3313 & .3324 & .1667 & .1696 \\
.3354 & .3346 & .1667 & .1634 \\
.3313 & .3314 & .1667 & .1706
\end{bmatrix}.
$$

Finally if we use as a starting vector $x_0 = (1,0,0,0)$, then from (2.4) we know that after $k = 10$ steps $x_0 P^{10}$ differs in norm from π by at most

$$\left(\tfrac{1}{3}\right)^{10} \cdot \left[\, \left\| \left(\tfrac{1}{2}, \tfrac{1}{2}, 0, -1\right) \right\| + \left\| \left(0, -\tfrac{5}{6}, 0, \tfrac{5}{6}\right) \right\| \,\right] \approx 6 \times 10^{-5}.$$

CHAPTER V

Nonstationary Markov Chains and the Ergodic Coefficient

In the previous chapters, we have studied stationary Markov chains using the classical approach and the algebraic approach. We have considered questions of reducibility, periodicity, and ergodicity (existence of a long run or invariant distribution) using these two approaches. In this chapter we will introduce a third approach to the study of ergodicity of stationary Markov chains, the ergodic coefficient approach. [The ergodic coefficient was defined by Dobrushin (1956). This approach has been considered by many authors since then. Some of the relevant papers and books include Hajnal (1958), Iosifescu (1966), Paz (1963, 1971), and Madsen and Isaacson (1973).] We will see that the applications to stationary Markov chains are rather simple corollaries of results that relate to nonstationary Markov chains. In view of this fact, our attention in this chapter will be devoted primarily to nonstationary Markov chains.

The reader may find that the level of mathematics used in this chapter is somewhat higher than in the previous chapters. By concentrating on the examples and the statements of the theorems rather than on the proofs of the theorems, readers who find the mathematics difficult should still be able to understand most of the concepts discussed in this chapter.

SECTION 1 **TYPES OF ERGODIC BEHAVIOR**

In Chapter I we discussed the difference between stationary and non-stationary Markov chains. Recall that for nonstationary chains, the transition matrix that gives the probabilities of going from state i at time $k-1$ to state j at time k, depends on k. We will denote this transition matrix by P_k or $P^{(k-1,k)}$.

A nonstationary Markov chain is completely described by a starting vector $\mathbf{f}^{(0)}$ and a sequence of transition matrices $\{P_k\}_{k=1}^{\infty}$. Recall that $\mathbf{f}^{(0)}$ is a starting vector if it gives a probability distribution over the states. In this chapter we will allow the state space to be countably infinite, hence the only restrictions on $\mathbf{f}^{(0)}$ are

$$f_i^{(0)} \geqslant 0, \qquad i = 1, 2, \ldots \quad \text{and} \quad \sum_{i=1}^{\infty} f_i^{(0)} = 1.$$

Definition V.1.1. *Let $P_1, P_2, \ldots$ be transition matrices for a nonstationary Markov chain with starting vector $\mathbf{f}^{(0)}$. We define*

$$\mathbf{f}^{(k)} = \mathbf{f}^{(0)} P_1 \cdot P_2 \cdot \ldots \cdot P_k \quad \text{and} \quad \mathbf{f}^{(m,k)} = \mathbf{f}^{(0)} P_{m+1} \cdot P_{m+2} \cdot \ldots \cdot P_k.$$

Define the jth element of $\mathbf{f}^{(k)}$ by $f_j^{(k)}$ and define the (i,j)th element of $P^{(m,k)} = P_{m+1} \cdot P_{m+2} \cdot \ldots \cdot P_k$ by $p_{ij}^{(m,k)}$. Similarly $\mathbf{g}^{(k)}$, $\mathbf{g}^{(m,k)}$, and $g_j^{(m,k)}$ will be used when $\mathbf{g}^{(0)}$ is the starting vector.

We are interested in studying the behavior of $\mathbf{f}^{(k)}$, or $\mathbf{f}^{(m,k)}$, as $k \to \infty$. It may be that $\mathbf{f}^{(k)}$ converges to the same fixed vector $\mathbf{q}$, no matter what starting vector $\mathbf{f}^{(0)}$ is used; that is, a limiting vector exists which is independent of choice of starting vector. (The reader will note that this is exactly what happens in the stationary case if the chain is ergodic.) In view of the fact that the limiting vector is independent of the starting vector, information about $\mathbf{f}^{(k)}$, for large k, gives us little or no information about $\mathbf{f}^{(0)}$; that is, the effect of the starting vector is lost after a long time. We will frequently refer to this loss of effect of $\mathbf{f}^{(0)}$ as *loss of memory*. When $\lim_{k \to \infty} \mathbf{f}^{(m,k)} = \mathbf{q}$ independently of $\mathbf{f}^{(0)}$, the behavior is referred to as *convergence and loss of memory*, and will be called strongly ergodic.

It is also possible to have convergence without loss of memory or loss of memory without convergence. In the latter case, the effect of the initial distribution is lost so that $\mathbf{f}^{(k)}$ and $\mathbf{g}^{(k)}$ are in some sense "close," but there is no convergence so that $\mathbf{f}^{(k)}$ and $\mathbf{f}^{(k+1)}$ need not be "close" even for very large k. Such behavior will be called weakly ergodic. The case of convergence without loss of memory is of less interest because of the dependence on the choice of starting vector and will not be discussed further here.

In order to give precise definitions of weakly and strongly ergodic behavior we must give a definition for the norm of a vector and the norm of a matrix. While there are many norms that could be placed upon spaces of vectors and matrices of countable dimension, we consider only one norm on each of these spaces. The symbol $\|\cdot\|$ will be used in both cases, with the meaning determined by whether there is a vector or a matrix inside the double bars.

Definition V.1.2. *If* $\mathbf{f} = (f_1, f_2, f_3, \ldots)$ *is a vector, define the norm of* $\mathbf{f}$ *by*

$$\|\mathbf{f}\| = \sum_{i=1}^{\infty} |f_i|.$$

If $A = (a_{ij})$ *is a square matrix, define the norm of* A *by*

$$\|A\| = \sup_i \sum_{j=1}^{\infty} |a_{ij}|.$$

(It is left as an exercise to show that these functions satisfy the conditions of a norm as given in Chapter I.)

Definition V.1.3. *A nonstationary Markov chain is called* weakly ergodic *if for all* m.

$$\lim_{k \to \infty} \sup_{\mathbf{f}^{(0)}, \mathbf{g}^{(0)}} \|\mathbf{f}^{(m,k)} - \mathbf{g}^{(m,k)}\| = 0, \tag{1.1}$$

where $\mathbf{f}^{(0)}$ *and* $\mathbf{g}^{(0)}$ *are starting vectors.*

Let us briefly consider a reason for including the phrase "for all m" in Definition V.1.3. Say we have a nonstationary Markov chain with starting vector $\mathbf{f}^{(0)}$ and transition matrices $P_1, P_2, P_3, \ldots$ corresponding to transitions at discrete times $t = 1, 2, 3, \ldots$. Clearly we could consider a "different" chain with transition times at $t_1^* = m+1, m+2, m+3, \ldots$ and transition matrices $P_1^* = P_{m+1}$, $P_2^* = P_{m+2}$, and so on. Although these chains are "different," they are intimately related. By giving the definition that we have, we are requiring that these "different" chains all exhibit the same kind of behavior, that is, no matter when the process starts, the chain will lose memory.

The following example further clarifies the need for the phrase "for all m" in Definition V.1.3.

Example V.1.1. Let $\{X_n\}$ be a nonstationary Markov chain with transition matrices

$$P_1 = \begin{pmatrix} \frac{1}{3} & \frac{2}{3} \\ \frac{1}{2} & \frac{1}{2} \end{pmatrix}, \qquad P_2 = \begin{pmatrix} \frac{1}{4} & \frac{3}{4} \\ \frac{1}{4} & \frac{3}{4} \end{pmatrix},$$

$$P_n = \begin{pmatrix} \dfrac{1}{n^2} & 1 - \dfrac{1}{n^2} \\ 1 - \dfrac{1}{n^2} & \dfrac{1}{n^2} \end{pmatrix}, \qquad n = 3, 4, \ldots.$$

Now

$$P_1 P_2 = \begin{pmatrix} \frac{1}{4} & \frac{3}{4} \\ \frac{1}{4} & \frac{3}{4} \end{pmatrix}, \qquad P_1 P_2 P_3 = \begin{pmatrix} \dfrac{25}{36} & \dfrac{11}{36} \\ \dfrac{25}{36} & \dfrac{11}{36} \end{pmatrix},$$

and in general

$$P_1 P_2 \cdots P_n = \begin{pmatrix} a_n & 1 - a_n \\ a_n & 1 - a_n \end{pmatrix}.$$

That is, $P_1 P_2 \cdots P_n$ is a constant matrix for all $n \geqslant 2$ simply because P_2 is a constant matrix. (We define a constant matrix to be a matrix that has identical rows.) For any starting vector $\mathbf{f}^{(0)} = (f_1^{(0)}, f_2^{(0)})$ we have $\mathbf{f}^{(n)} = f^{(0)} P_1 P_2 \cdots P_n = (a_n, 1 - a_n)$ independently of $\mathbf{f}^{(0)}$, and so the process seems to be weakly ergodic since memory is lost. However, closer examination reveals that all memory is lost at step two (one might say that the chain suffers instant amnesia at the second step). We would prefer to have loss of memory depend on the tail of the nonstationary chain, and so we require that memory be lost independently of when the process starts. (The reader may recall that a similar "tail-dependent" requirement is used in defining divergence of an infinite product. In that case if a sequence contains a finite number of zeroes, it does not necessarily follow that the corresponding infinite product diverges to zero.) In fact the chain in this example is not weakly ergodic. (Exercise 1 of Section V.3.)

We would like to point out here that other authors have found it useful to define "weak ergodicity" differently than we have. For example, the condition that convergence be uniform over $\mathbf{f}^{(0)}$ and $\mathbf{g}^{(0)}$ could be dropped or the condition that convergence be uniform over m could be added. [For example, see Griffeath (1974) or Iosifescu and Theodorescu (1969).] In this book we restrict our attention to weak ergodicity as defined in Definition V.1.3 with the exception of an occasional remark or problem referring to some other type of weak ergodicity.

Definition V.1.4. *A nonstationary Markov chain is called* strongly ergodic *if there exists a vector* $\mathbf{q} = (q_1, q_2, \ldots)$, *with* $\|\mathbf{q}\| = 1$ *and* $q_i \geqslant 0$, *for* $i = 1, 2, 3, \ldots$ *such that for all* m

$$\lim_{k \to \infty} \sup_{\mathbf{f}^{(0)}} \|\mathbf{f}^{(m,k)} - \mathbf{q}\| = 0, \tag{1.2}$$

where $\mathbf{f}^{(0)}$ *is a starting vector.*

It is left to the reader (Exercise 5 of Section V.2) to show that strong ergodicity implies weak ergodicity.

At this point we should point out that the mode of convergence to zero used to define weak and strong ergodicity is stronger than coordinatewise convergence. Referring to Definition V.1.2, we see that it is not sufficient to have $\lim_{n\to\infty} |f_i^{(m,n)} - g_i^{(m,n)}| = 0$ or $\lim_{n\to\infty} |f_i^{(m,n)} - q_i| = 0$ (i.e., coordinatewise convergence to zero) in order to have weak or strong ergodicity, rather it is necessary to have

$$\lim_{n\to\infty} \sum_{i=1}^{\infty} |f_i^{(m,n)} - g_i^{(m,n)}| = 0 \quad \text{or} \quad \lim_{n\to\infty} \sum_{i=1}^{\infty} |f_i^{(m,n)} - q_i| = 0.$$

For finite chains these two types of convergence are equivalent, but this is not the case for infinite chains. In the remainder of this chapter we will make extensive use of the fact that convergence using the norm of Definition V.1.2 for infinite chains is a stronger mode of convergence than coordinatewise convergence. The following example illustrates this point.

Example V.1.2. Let $\mathbf{f}^{(n)} = (1/n, 1/n, \ldots 1/n, 0, 0, \ldots)$, where the first n coordinates of $\mathbf{f}^{(n)}$ are $1/n$. Since $\lim_{n\to\infty} f_i^{(n)} = 0$ for all i, the limit of $\mathbf{f}^{(n)}$ is the zero vector. However, $\|\mathbf{f}^{(n)} - \boldsymbol{\varphi}\| = 1$ for all n, so $\{\mathbf{f}^{(n)}\}$ does not converge in norm to the zero vector.

It is worth noting that if $\{\mathbf{f}^{(n)}\}$ is a sequence of probability vectors converging in norm to a vector $\mathbf{q}$, say, then $\mathbf{q}$ is also a probability vector, that is, $q_i \geqslant 0$ for all i and $\sum_{i=1}^{\infty} q_i = 1$ (Exercise 5).

It is, of course, generally quite difficult to show directly from the definitions that a chain is weakly or strongly ergodic. The main difficulty stems from the fact that one must consider the supremum over all possible starting vectors. The following examples are contrived to get around this difficulty, but do illustrate these two types of ergodic behavior.

Example V.1.3. Let $\{X_n\}$ be a nonstationary Markov chain with transition matrices

$$P_{2n-1} = \begin{bmatrix} 1 - \dfrac{1}{2n-1} & \dfrac{1}{2n-1} \\ 1 - \dfrac{1}{2n-1} & \dfrac{1}{2n-1} \end{bmatrix}, \quad P_{2n} = \begin{bmatrix} \dfrac{1}{2n} & 1 - \dfrac{1}{2n} \\ \dfrac{1}{2n} & 1 - \dfrac{1}{2n} \end{bmatrix}$$

for $n = 1, 2, \ldots$. Then, for any starting vector $\mathbf{f}^{(0)}$, we have

$$\mathbf{f}^{(m,k)} = \left(1 - \frac{1}{k}, \frac{1}{k}\right) \quad \text{if } k \text{ is odd}$$

$$= \left(\frac{1}{k}, 1 - \frac{1}{k}\right) \quad \text{if } k \text{ is even,}$$

so the chain is weakly ergodic but not strongly ergodic.

Example V.1.4. Let $\{X_n\}$ be a nonstationary Markov chain with transition matrices

$$P_n = \begin{bmatrix} \dfrac{1}{2} - \dfrac{1}{n+1} & \dfrac{1}{2} + \dfrac{1}{n+1} \\ \dfrac{1}{2} - \dfrac{1}{n+1} & \dfrac{1}{2} + \dfrac{1}{n+1} \end{bmatrix}$$

for $n = 1, 2, \ldots$. Then, for any starting vector $\mathbf{f}^{(0)}$, we have

$$\mathbf{f}^{(m,k)} = \left(\frac{1}{2} - \frac{1}{k+1}, \frac{1}{2} + \frac{1}{k+1}\right) \to \left(\frac{1}{2}, \frac{1}{2}\right), \quad \text{as } k \to \infty$$

so the chain is strongly ergodic.

EXERCISES

1. Let $\mathbf{f} = (1, -\frac{1}{2}, \frac{1}{4}, -\frac{1}{8}, \ldots)$ and let

$$A = \begin{bmatrix} 1 & \frac{1}{2} & \frac{1}{4} & \frac{1}{8} & \cdots \\ \frac{1}{2} & \frac{1}{4} & \frac{1}{8} & \frac{1}{16} & \cdots \\ \frac{1}{4} & \frac{1}{8} & \frac{1}{16} & \frac{1}{32} & \cdots \\ \vdots & \vdots & \vdots & \vdots & \end{bmatrix}.$$

   Find $\|\mathbf{f}\|$, $\|A\|$, and $\|\mathbf{f}A\|$.

2. Show that the two norms given in Definition V.1.2 satisfy the properties of a norm as given in Chapter I.

3. If A is a matrix, define a norm for A by

$$|A| = \sup_{i,j} |a_{ij}|.$$

(a)   Show that this function satisfies the properties of a norm as given in Chapter I.

(b)   Let $\{A_n\}$ be a sequence of matrices. Show that $\|A_n\|\to 0$ as $n\to\infty$ implies that $|A_n|\to 0$ as $n\to\infty$.

(c)   Show that $|A_n|\to 0$ as $n\to\infty$ implies elementwise convergence of $\{A_n\}$ to zero. [Note that for finite dimensional matrices the three modes of convergence referred to in (b) and (c) are equivalent in the sense that if $\{A_n\}$ converges to zero in one of the norms, it converges to zero all three norms.]

(d)   Give an example of a sequence of matrices $\{A_n\}$ such that $\lim_{n\to\infty}|A_n|=0$ but $\lim_{n\to\infty}\|A_n\|\neq 0$.

(e)   Give an example of a sequence of matrices $\{A_n\}$ such that $\lim_{n\to\infty}A_n=0$ elementwise but $\lim_{n\to\infty}|A_n|\neq 0$.

4.   Find $\|A\|$ and $|A|$ (as defined in the previous problem) for

$$\text{(a)}\qquad A=\begin{bmatrix} \frac{1}{3} & \frac{1}{6} & \frac{1}{2} \\ \frac{1}{5} & \frac{2}{5} & \frac{2}{5} \\ \frac{1}{4} & \frac{1}{2} & \frac{1}{4} \end{bmatrix}.$$

$$\text{(b)}\qquad A=\begin{bmatrix} \frac{1}{3} & \frac{1}{6} & -\frac{1}{2} \\ \frac{1}{3} & \frac{1}{3} & -\frac{2}{3} \\ \frac{1}{4} & -\frac{1}{2} & \frac{1}{4} \end{bmatrix}.$$

5.   Using the norm given in Definition V.1.2, prove that if $\{\mathbf{f}_n\}_{n=1}^{\infty}$ is a sequence of probability vectors converging in norm to a vector $\mathbf{q}$, then $\mathbf{q}$ is a probability vector.

6.   Let $\{P_k\}$ be a sequence of stochastic matrices defining a nonstationary Markov chain. If $\{P_k\}$ contains an infinite number of constant matrices, explain why the corresponding chain must be weakly ergodic.

SECTION 2   **THE ERGODIC COEFFICIENT**

In Section V.1 we indicated that it is difficult to show that a chain is weakly or strongly ergodic by using the definitions directly. In view of this difficulty we introduce the ergodic coefficient as defined by Dobrushin (1956). (In reviewing the literature it is clear that other authors have defined functions of stochastic matrices which play a role similar to the ergodic coefficient of Dobrushin in determining the ergodic behavior of stationary and nonstationary Markov chains. However, since Dobrushin's paper makes extensive use of the ergodic coefficient and discusses its properties in a very general case, we have given that paper as the main reference.) The ergodic coefficient can be used to give characterizations of weak and strong ergodicity that are somewhat easier to check than the definitions themselves. After defining the ergodic coefficient, we give some lemmas that describe some of its important properties.

Definition V.2.1. *Let P be a stochastic matrix. The* ergodic coefficient *of P, denoted by* $\alpha(P)$, *is defined by*

$$\alpha(P) = 1 - \sup_{i,k} \sum_{j=1}^{\infty} \left[p_{ij} - p_{kj} \right]^{+} \tag{2.1}$$

where $[p_{ij} - p_{kj}]^{+} = max(0, p_{ij} - p_{kj})$.

Note that this definition is given for a matrix with countably many rows and columns; that is, for the case where the state space is countable. Actually, Dobrushin (1956) defined the ergodic coefficient for the case where the state space is continuous. However, in this chapter, we give results only for the case where $S = \{1, 2, 3, \ldots\}$ and thereby cover the cases of finite state space and countably infinite state space. When results extend to a more general state space, this will be indicated.

Lemma V.2.1. The ergodic coefficient satisfies the following conditions:

(i) $0 \leqslant \alpha(P) \leqslant 1$

(ii) $\alpha(P) = 1 - (\tfrac{1}{2}) \sup_{i,k} \sum_{j=1}^{\infty} |p_{ij} - p_{kj}|$.

Proof. The proof is straightforward and is left as an exercise. ▲

Lemma V.2.2. Let P be a stochastic matrix. Then

$$\alpha(P) = \inf_{i,k} \sum_{j=1}^{\infty} \min(p_{ij}, p_{kj}).$$

Proof. Let i and k be fixed. Since $(p_{ij} - p_{kj})^{+} = [p_{ij} - \min(p_{ij}, p_{kj})]$ and since $\sum_{j=1}^{\infty} p_{ij} = 1$, we have

$$1 - \sum_{j=1}^{\infty} (p_{ij} - p_{kj})^{+} = 1 - \sum_{j=1}^{\infty} (p_{ij} - \min(p_{ij}, p_{kj}))$$

$$= \sum_{j=1}^{\infty} \min(p_{ij}, p_{kj}).$$

Taking the infimum of both sides over i and k, we get

$$\inf_{i,k} \sum_{j=1}^{\infty} \min(p_{ij}, p_{kj}) = \inf_{i,k} \left[1 - \sum_{j=1}^{\infty} (p_{ij} - p_{kj})^+ \right]$$

$$= 1 - \sup_{i,k} \sum_{j=1}^{\infty} (p_{ij} - p_{kj})^+. \quad \blacktriangle$$

It sometimes more convenient to use $1 - \alpha(P)$ instead of $\alpha(P)$ itself. In view of this we define

$$\delta(P) = 1 - \alpha(P)$$

and call $\delta(P)$ the *delta coefficient* of P. It follows from Definition V.2.1 and Lemmas V.2.1 and V.2.2 that

$$\delta(P) = \sup_{i,k} \sum_{j=1}^{\infty} \left[p_{ij} - p_{kj} \right]^+ = \left(\tfrac{1}{2} \right) \sup_{i,k} \sum_{j=1}^{\infty} |p_{ij} - p_{kj}|$$

$$= 1 - \inf_{i,k} \sum_{j=1}^{\infty} \min(p_{ij}, p_{kj}).$$

At this point two non-negative functions have been defined on subsets of the set of countable matrices. The norm, $\| \cdot \|$, has been defined on all such matrices and the delta coefficient, $\delta(\cdot)$, has been defined on the subset of stochastic matrices. The following discussion is intended to give the reader a feeling for what these functions are measuring. The norm is not used to measure a property of a stochastic matrix since the norm of a stochastic matrix is always one. In this book the norm will usually be applied to the difference of two stochastic matrices and will be measuring how "close" the matrices are to each other. The norm of a matrix, A, is zero if and only if all the entries of A are zero. Recall that for finite matrices $\|A_n\| \to 0$ if and only if the entries of A_n converge to zero as n goes to infinity. However, for infinite matrices convergence to zero in the norm, $\| \cdot \|$, is stronger than elementwise convergence. (See Exercise 3 of Section V.1).

The delta coefficient, $\delta(\cdot)$, is defined on stochastic matrices in this chapter and will later be extended to non-negative matrices. We can see that δ is not used to measure how close a matrix is to zero but rather to measure how close a non-negative matrix is to being a constant matrix. In fact, $\delta(P) = 0$ if and only if P is a constant matrix. In Section V.3 we will see that $\{P_n\}$ is weakly ergodic if and only if $P_m P_{m+1} \cdots P_n$ is close to a constant matrix for large n and δ is used to measure this closeness.

We conclude this section with three important inequalities involving $\delta(\cdot)$ and $\|\cdot\|$. These inequalities along with the triangle inequality for the norm are basic tools in proving theorems in this chapter. (A comment on notation is in order at this point. In Section III.3 we used the letter Q to represent a substochastic matrix. In this chapter the letter Q will be used to represent either an ordinary stochastic matrix or a constant stochastic matrix. It should be clear from the context when Q is intended to be a constant stochastic matrix.)

Lemma V.2.3.   If P and Q are stochastic matrices, then

$$\delta(QP) \leqslant \delta(Q)\delta(P).$$

Proof.   In Definition V.2.1 we introduced the notation $a^+ = \max(0, a)$. If we also use a^- to denote $\max(0, -a)$, then we have $a = a^+ - a^-$. Using this notation we see that for any two rows i and k of a stochastic matrix Q we have

$$\sum_{j=1}^{\infty} (q_{ij} - q_{kj})^+ = \sum_{j=1}^{\infty} (q_{ij} - q_{kj})^-.$$

This is true since

$$\sum_{j=1}^{\infty} (q_{ij} - q_{kj})^+ - \sum_{j=1}^{\infty} (q_{ij} - q_{kj})^- = \sum_{j=1}^{\infty} (q_{ij} - q_{kj}) = 1 - 1 = 0.$$

If we define $QP = R = (r_{ij})$, then $\delta(QP) = \delta(R) = \sup_{i,k} \sum_{l=1}^{\infty} (r_{il} - r_{kl})^+$. For the moment fix i and k and consider

$$\sum_{l=1}^{\infty} (r_{il} - r_{kl})^+ = \sum_{l=1}^{\infty} \left[\sum_{j} q_{ij}p_{jl} - q_{kj}p_{jl} \right]^+. \tag{2.2}$$

Let $E = \{l : \sum_{j=1}^{\infty} (q_{ij} - q_{kj}) p_{jl} > 0\}$. That is, E denotes those columns, l, for which the value of $r_{il} - r_{kl}$ is positive. Using this set E, (2.2) can be written as

$$\sum_{l \in E} \sum_{j=1}^{\infty} (q_{ij} - q_{kj}) p_{jl}. \tag{2.3}$$

The order of summation can be interchanged using Fubini's theorem (Exercise 4) so (2.3) is equal to

$$\sum_{j=1}^{\infty}(q_{ij}-q_{kj})\sum_{l\in E}p_{jl}=\sum_{j=1}^{\infty}\left[(q_{ij}-q_{kj})^{+}-(q_{ij}-q_{kj})^{-}\right]\sum_{l\in E}p_{jl}$$

$$=\sum_{j=1}^{\infty}(q_{ij}-q_{kj})^{+}\sum_{l\in E}p_{jl}-\sum_{j=1}^{\infty}(q_{ij}-q_{kj})^{-}\sum_{l\in E}p_{jl}.$$

Now since all the terms in this difference are non-negative, the difference is made larger if the first term is increased and the second is decreased. That is, in place of $\sum_{l\in E}p_{jl}$ we substitute $\sup_{j}\sum_{l\in E}p_{jl}$ in the first term of the difference and $\inf_{j}\sum_{l\in E}p_{jl}$ in the second term.

Using the fact that

$$\sum_{j=1}^{\infty}(q_{ij}-q_{kj})^{+}=\sum_{j=1}^{\infty}(q_{ij}-q_{kj})^{-}$$

we get

$$\sum_{l=1}^{\infty}(r_{il}-r_{kl})^{+}\leqslant\sum_{j=1}^{\infty}(q_{ij}-q_{kj})^{+}\left[\sup_{j}\sum_{l\in E}p_{jl}-\inf_{j}\sum_{l\in E}p_{jl}\right]$$

$$=\sum_{j=1}^{\infty}(q_{ij}-q_{kj})^{+}\sup_{j_{1}j_{2}}\sum_{l\in E}(p_{j_{1}l}-p_{j_{2}l})$$

$$\leqslant\sum_{j=1}^{\infty}(q_{ij}-q_{kj})^{+}\sup_{j_{1}j_{2}}\sum_{l=1}^{\infty}(p_{j_{1}l}-p_{j_{2}l})^{+}.$$

[Note that in the last expression the summation is taken over all l but no negative values are introduced since $(p_{j_{1}l}-p_{j_{2}l})$ was changed to $(p_{j_{1}l}-p_{j_{2}l})^{+}$.]

The last expression is simplified to $\sum_{j=1}^{\infty}(q_{ij}-q_{kj})^{+}\delta(P)$. Hence we have

$$\sum_{l=1}^{\infty}(r_{il}-r_{kl})^{+}\leqslant\sum_{j=1}^{\infty}(q_{ij}-q_{kj})^{+}\delta(P),$$

so taking the supremums of both sides over i and k we get $\delta(QP)\leqslant\delta(Q)\delta(P)$. Note that in terms of the ergodic coefficient this lemma says $1-\alpha(QP)\leqslant[1-\alpha(Q)][1-\alpha(P)]$, which is the form of the inequality in Dobrushin (1956). ▲

Lemma V.2.4. If P is a stochastic matrix and if R is any matrix with $\sum_{k=1}^{\infty} r_{ik} = 0$ for all i and $\|R\| < \infty$, then $\|RP\| \leqslant \|R\| \delta(P)$.

Proof. The main ideas of the proof follow very closely the proof of the previous lemma so the details are not repeated here. It is sufficient to note that the r_{ij} of this lemma play the role of the $(q_{ij} - q_{kj})$ of the previous lemma. Also since $\sum_{k=1}^{\infty} (\sum_{j=1}^{\infty} r_{ij} p_{jk}) = 0$ for all i we have

$$\|RP\| = 2 \sup_i \sum_{k=1}^{\infty} \left(\sum_{j=1}^{\infty} r_{ij} p_{jk} \right)^+ . \quad \blacktriangle$$

We would like to emphasize that the ordering of the matrices R and P in Lemma V.2.4 is crucial. The following example shows that if the matrix, R, with rows summing to zero is not on the left, then the inequality may fail.

Example V.2.1. Let the matrices P and R be defined as follows:

$$P = \begin{pmatrix} 1 & 0 \\ 1 & 0 \end{pmatrix}, \qquad R = \begin{pmatrix} 1 & -1 \\ 0 & 0 \end{pmatrix}$$

Then

$$PR = \begin{pmatrix} 1 & -1 \\ 1 & -1 \end{pmatrix},$$

so that $\|PR\| \not\leqslant \delta(P)\|R\|$.

There are several instances throughout this chapter where one might be tempted to use an incorrect form of Lemma V.2.4. Therefore Example V.2.1 should be kept in mind as a reminder of the importance of the ordering of R and P.

Lemma V.2.5. For all matrices A and B, the following inequality holds:

$$\|AB\| \leqslant \|A\| \cdot \|B\|$$

Proof. The case where $\|A\|$ or $\|B\|$ is either zero or infinite, is easily done and left to the reader as an exercise. Therefore assume that $0 < \|A\| < \infty$ and $0 < \|B\| < \infty$. Note that the (i,j)th element of AB is given by $\sum_{k=1}^{\infty} a_{ik} b_{kj}$. Then

$$\|AB\| = \sup_i \sum_{j=1}^{\infty} \left| \sum_{k=1}^{\infty} a_{ik} b_{kj} \right|$$

$$\leqslant \sup_i \sum_{j=1}^{\infty} \sum_{k=1}^{\infty} |a_{ik}| |b_{kj}|.$$

By Fubini's theorem, this last expression is equal to

$$\sup_i \sum_{k=1}^{\infty} |a_{ik}| \sum_{j=1}^{\infty} |b_{kj}|$$

$$\leqslant \sup_i \sum_{k=1}^{\infty} |a_{ik}| \cdot \sup_k \sum_{j=1}^{\infty} |b_{kj}|$$

$$= \|A\| \cdot \|B\|. \quad \blacktriangle$$

EXERCISES

1. Find $\delta(P)$, $\delta(Q)$, $\delta(PQ)$, $\delta(QP)$, and $\|Q - P\|$ where

   (a) $\quad P = \begin{bmatrix} .1 & .3 & .6 \\ 0 & .5 & .5 \\ .4 & .6 & 0 \end{bmatrix}$ and $Q = \begin{bmatrix} 0 & .4 & .6 \\ .5 & .5 & 0 \\ .3 & .4 & .3 \end{bmatrix}$,

   (b) $\quad P = \begin{pmatrix} .3 & .7 \\ .5 & .5 \end{pmatrix}$ and $Q = \begin{pmatrix} .4 & .6 \\ 1 & 0 \end{pmatrix}$.

2. Find $\|RP\|$, $\|R\| \cdot \delta(P)$, and $\|PR\|$ where

   $$R = \begin{bmatrix} 1 & 0 & -1 \\ \frac{1}{2} & \frac{1}{2} & -1 \\ 1 & -1 & 0 \end{bmatrix} \quad \text{and} \quad P = \begin{bmatrix} \frac{1}{2} & 0 & \frac{1}{2} \\ 1 & 0 & 0 \\ \frac{1}{4} & \frac{1}{2} & \frac{1}{4} \end{bmatrix}.$$

   Relate your answers to this problem to Lemma V.2.4 and the ensuing discussion.

3. Find $\|A\|$, $\|B\|$, $\|AB\|$, and $\|BA\|$ where

   $$A = \begin{bmatrix} 1 & -1 & 1 \\ 2 & 0 & -1 \\ 1 & 0 & 0 \end{bmatrix} \quad \text{and} \quad B = \begin{bmatrix} \frac{1}{2} & 0 & 1 \\ -1 & 0 & -2 \\ \frac{1}{3} & 1 & -1 \end{bmatrix}.$$

4. Justify the use of Fubini's theorem in the proof of Lemma V.2.3.
5. Show that strong ergodicity implies weak ergodicity.
6. Prove Lemma V.2.1.
7. Give a detailed proof of Lemma V.2.4.

SECTION 3   **WEAK ERGODICITY**

In this section we give several theorems in which the ergodic coefficient can be used to determine whether a nonstationary Markov chain is weakly ergodic. In many cases one of these theorems will be easier to use than the original definition.

Theorem V.3.1.

A nonstationary Markov chain is weakly ergodic if and only if for all m, $\delta(P^{(m,k)}) \to 0$ as $k \to \infty$.

Proof. Assume that for all m, $\delta(P^{(m,k)}) \to 0$ as $k \to \infty$. Let $\mathbf{f}^{(0)}$ and $\mathbf{g}^{(0)}$ be any two starting vectors and let m and k be fixed. Define a stochastic matrix Q such that the first row is $\mathbf{f}^{(0)}$ and the remaining rows are $\mathbf{g}^{(0)}$. Consider the matrix $QP^{(m,k)} = R$. The first row of the matrix R is $\mathbf{f}^{(m,k)}$ and the remaining rows are $\mathbf{g}^{(m,k)}$. Therefore, since the value of $\delta(R)$ is determined by the rows of R, we have

$$\delta(QP^{(m,k)}) = \delta(R) = \frac{1}{2} \sup_{i,j} \sum_{l=1}^{\infty} |r_{il} - r_{jl}| = \frac{1}{2} \sum_{l=1}^{\infty} |f_l^{(m,k)} - g_l^{(m,k)}|$$

$$= \frac{1}{2} \|\mathbf{f}^{(m,k)} - \mathbf{g}^{(m,k)}\|.$$

Using Lemma V.2.3 and the fact that $\delta(Q) \leqslant 1$, we see that

$$\|\mathbf{f}^{(m,k)} - \mathbf{g}^{(m,k)}\| = 2\delta(QP^{(m,k)}) \leqslant 2\delta(Q)\delta(P^{(m,k)}) \leqslant 2\delta(P^{(m,k)}).$$

By assumption, the right-hand side goes to zero for each m as $k \to \infty$. Furthermore, it goes to zero independently of $\mathbf{f}^{(0)}$ and $\mathbf{g}^{(0)}$, so the chain is weakly ergodic.

Conversely, assume that for all m, $\sup_{\mathbf{f}^{(0)}, \mathbf{g}^{(0)}} \|\mathbf{f}^{(m,k)} - \mathbf{g}^{(m,k)}\| \to 0$ as $k \to \infty$. Define $\hat{\mathbf{f}}^{(0)}$ to be a starting vector with a one in the ith position and zeros elsewhere and define $\hat{\mathbf{g}}^{(0)}$ to be a starting vector with a one in the jth position and zeros elsewhere. Note that the vectors $\hat{\mathbf{f}}^{(0)}P^{(m,k)} = \hat{\mathbf{f}}^{(m,k)}$ and $\hat{\mathbf{g}}^{(0)}P^{(m,k)} = \hat{\mathbf{g}}^{(m,k)}$ are the ith and jth rows of $P^{(m,k)}$, respectively, so

$$\sum_{l=1}^{\infty} |p_{il}^{(m,k)} - p_{jl}^{(m,k)}| = \|\hat{\mathbf{f}}^{(m,k)} - \hat{\mathbf{g}}^{(m,k)}\| \leqslant \sup_{\mathbf{f}^{(0)}, \mathbf{g}^{(0)}} \|\mathbf{f}^{(m,k)} - \mathbf{g}^{(m,k)}\|.$$

Since this inequality holds for all i and j, it follows that

$$2\delta(P^{(m,k)}) = \sup_{i,j} \sum_{l=1}^{\infty} |p_{il}^{(m,k)} - p_{jl}^{(m,k)}| \leqslant \sup_{\mathbf{f}^{(0)}, \mathbf{g}^{(0)}} \|\mathbf{f}^{(m,k)} - \mathbf{g}^{(m,k)}\|,$$

and this last term tends to zero for all m as $k \to \infty$ by assumption. ▲

In using Theorem V.3.1 to establish weak ergodicity, it is sometimes convenient to also use Lemma V.2.3. That is, since $\delta(P^{(m,k)}) \leqslant \Pi_{j=m+1}^{k} \delta(P_j)$, it is sufficient to show that this product goes to zero for all m. As one might expect, this simplification is too crude in some situations, as can be seen by the following example.

Example V.3.1. Let $\{X_n\}$ be a nonstationary Markov chain with transition matrices

$$P_{2n-1} = \begin{bmatrix} 0 & \frac{1}{2} & \frac{1}{2} \\ 1 & 0 & 0 \\ 1 & 0 & 0 \end{bmatrix} \quad \text{and} \quad P_{2n} = \begin{bmatrix} 0 & 1 & 0 \\ \frac{1}{2} & 0 & \frac{1}{2} \\ 0 & 1 & 0 \end{bmatrix}$$

for $n = 1, 2, 3, \dots$.

It is easy to see that $\delta(P_{2n-1}) = 1$ and $\delta(P_{2n}) = 1$. Using

$$\delta(P^{(m,k)}) \leqslant \prod_{j=m+1}^{k} \delta(P_j) = 1 \cdot 1 \cdots \cdot 1 = 1,$$

we can say nothing about weak ergodicity. However,

$$P_1 P_2 = P_3 P_4 = \cdots = P_{2n-1} P_{2n} = \begin{bmatrix} 0 & \frac{1}{2} & \frac{1}{2} \\ 1 & 0 & 0 \\ 1 & 0 & 0 \end{bmatrix} \begin{bmatrix} 0 & 1 & 0 \\ \frac{1}{2} & 0 & \frac{1}{2} \\ 0 & 1 & 0 \end{bmatrix}$$

$$= \begin{bmatrix} \frac{1}{4} & \frac{1}{2} & \frac{1}{4} \\ 0 & 1 & 0 \\ 0 & 1 & 0 \end{bmatrix},$$

so $\delta(P_{2n-1} P_{2n}) = \frac{1}{2}$. Now if m is even and k is even,

$$\delta(P^{(m,k)}) \leqslant \delta(P_{m+1} \cdot P_{m+2}) \delta(P_{m+3} \cdot P_{m+4}) \cdots \cdot \delta(P_{k-1} \cdot P_k) = \left(\frac{1}{2}\right)^{(k-m)/2}.$$

The reader can show that for m even and k odd or m odd and k either even or odd, $\delta(P^{(m,k)}) \leqslant \left(\frac{1}{2}\right)^{[(k-m)/2]-1}$, where $[\cdot]$ represents the greatest integer function. Consequently the chain is weakly ergodic even though $\Pi_{j=m+1}^{k} \delta(P_j) = 1$ for all m and k.

We see from Example V.3.1 that a useable bound on $\delta(P^{(m,k)})$ can sometimes be obtained by finding the delta coefficient of "blocks" of products of transition matrices. This technique is exploited in the next theorem.

Theorem V.3.2.

Let $\{X_n\}$ be a nonstationary Markov chain with transition matrices $\{P_n\}_{n=1}^{\infty}$. The chain $\{X_n\}$ is weakly ergodic if and only if there exists a subdivision of $P_1 \cdot P_2 \cdot P_3 \cdots$ into blocks of matrices $[P_1 \cdot P_2 \cdots \cdot P_{n_1}] \cdot [P_{n_1+1} \cdot P_{n_1+2} \cdots \cdot P_{n_2}] \cdots [P_{n_j+1} \cdot P_{n_j+2} \cdots \cdot P_{n_{j+1}}] \ldots$ such that

$$\sum_{j=0}^{\infty} \alpha(P^{(n_j, n_{j+1})}) = \infty,$$

where $n_0 = 0$.

Proof. The first part of the proof depends on the following result from analysis: If $\{\epsilon_j\}_{j=0}^{\infty}$ is a sequence of numbers with $0 < \epsilon_j < 1$ for all j, then the product $\Pi_{j=m}^{n}(1 - \epsilon_j)$ diverges to zero as $n \to \infty$ if and only if $\Sigma_{j=m}^{\infty} \epsilon_j = \infty$ (see Section I.4). If $\Sigma_{j=0}^{\infty} \alpha(P^{(n_j, n_{j+1})}) = \infty$, then $\Sigma_{j=i}^{\infty} \alpha(P^{(n_j, n_{j+1})}) = \infty$ for all i. Using $\delta(P) = 1 - \alpha(P)$ we see that

$$\prod_{j=i}^{l} \delta(P^{(n_j, n_{j+1})}) = \prod_{j=i}^{l} \left[1 - \alpha(P^{(n_j, n_{j+1})})\right] \to 0 \quad \text{as } l \to \infty. \qquad (3.1)$$

Finally, let m be given and define $i = \min\{j : n_j > m\}$ and for $k > m$, define $l = \max\{j : n_j < k\}$ and note that $l \to \infty$ as $k \to \infty$. Then using (3.1) and Lemma V.2.3 we have

$$\delta(P^{(m,k)}) \leqslant \delta(P^{(m,n_i)}) \prod_{j=i}^{l-1} \delta(P^{(n_j, n_{j+1})}) \delta(P^{(n_l, k)}) \to 0 \quad \text{as} \quad k \to \infty.$$

Conversely, assume that the chain is weakly ergodic, that is for all m, $\delta(P^{(m,k)}) \to 0$ as $k \to \infty$. This implies that for all m, $\alpha(P^{(m,k)}) \to 1$ as $k \to \infty$. Hence for $m = 0 = n_0$, there exists n_1 such that $\alpha(P^{(0,n_1)}) > \frac{1}{2}$. Likewise, given n_1 there exists $n_2 > n_1$ such that $\alpha(P^{(n_1,n_2)}) > \frac{1}{2}$. Continuing in this manner we get

$$\sum_{j=0}^{k} \alpha(P^{(n_j, n_{j+1})}) > \frac{k+1}{2},$$

which diverges as $k \to \infty$. Hence we have constructed a partition of the original sequence of matrices $P_1, P_2, \ldots$ into blocks satisfying $\Sigma_{j=0}^{\infty} \alpha(P^{(n_j, n_{j+1})}) = \infty$. ▲

There are several other characterizations of weak ergodicity. We will only state two more, the proofs of which are contained in Paz (1971). (Recall that a constant stochastic matrix is a stochastic matrix with identical rows.)

Theorem V.3.3.

A nonstationary Markov chain is weakly ergodic if and only if for each m there is a sequence of constant stochastic matrices, Q_{mk} such that $\lim_{k \to \infty} \| P^{(m,k)} - Q_{mk} \| = 0$.

Theorem V.3.4.

Let $\{P_n\}_{n=1}^{\infty}$ be the transition matrices for a nonstationary Markov chain. Let P_n be written as $P_n = Q_n + R_n$ where Q_n is a constant stochastic matrix. Then $\{P_n\}_{n=1}^{\infty}$ is weakly ergodic if and only if for all m, $\lim_{k \to \infty} \| \Pi_{n=m}^{k+m} R_n \| = 0$.

The following is an example of a nonstationary Markov chain that is weakly ergodic but not strongly ergodic. We illustrate the use of each of the four theorems in this section by using them to show that the chain is weakly ergodic.

Example V.3.2.　Let $\{X_n\}$ be a nonstationary Markov chain with transition matrices

$$P_{2n-1} = \begin{pmatrix} \frac{1}{2} & \frac{1}{2} \\ 1 & 0 \end{pmatrix} \quad \text{and} \quad P_{2n} = \begin{pmatrix} 0 & 1 \\ 1 & 0 \end{pmatrix}$$

for $n = 1, 2, 3, \ldots$. To see that the chain is not strongly ergodic, let $\mathbf{f}^{(0)} = (0, 1)$. Then

$$\mathbf{f}^{(2k)} = \mathbf{f}^{(0)}(P_1 P_2) \cdot (P_3 \cdot P_4) \cdots (P_{2k-1} P_{2k})$$

$$= (0, 1) \begin{pmatrix} \frac{1}{2} & \frac{1}{2} \\ 0 & 1 \end{pmatrix}^k = (0, 1)$$

and

$$\mathbf{f}^{(2k+1)} = \mathbf{f}^{(0)}(P_1 P_2) \cdots (P_{2k-1} P_{2k}) P_{2k+1}$$

$$= (0, 1) \begin{pmatrix} \frac{1}{2} & \frac{1}{2} \\ 1 & 0 \end{pmatrix} = (1, 0).$$

Hence we have $\|\mathbf{f}^{(2j)} - \mathbf{f}^{(2j+1)}\| = 2$ for $j = k, k+1, \ldots$, and for this particular starting vector the sequence $\{\mathbf{f}^{(k)}\}$ does not converge, so the chain is not strongly ergodic. We can however show that the chain is weakly ergodic in several different ways.

(i)   Use Theorem V.3.1 to show weak ergodicity. Since $\delta(P_{2n}) = 1$ and $\delta(P_{2n-1}) = \frac{1}{2}$, it follows that

$$\delta(P^{(m,k)}) \leqslant \prod_{j=m+1}^{k} \delta(P_j) \leqslant (\tfrac{1}{2})^{[(k-m)/2]} \to 0 \quad \text{as } k \to \infty$$

for all m. (Here again $[\cdot]$ represents the greatest integer function.)

(ii)   Use Theorem V.3.2. Partition $\{P_n\}_{n=1}^{\infty}$ into blocks consisting of two matrices each. That is partition $P_1 P_2 P_3 \ldots$ into $(P_1 P_2)(P_3 P_4) \ldots$. Since $\alpha(P_j P_{j+1}) = \frac{1}{2}$ for all j, we have $\sum_{j=1}^{\infty} \alpha(P_{2j-1} P_{2j}) = \infty$.

In this particular example, blocks of size 1 could have been used just as easily to get a divergent sum.

(iii)   Use Theorem V.3.3. If m is even and k $(k > m)$ is even, then

$$P^{(m,k)} = \begin{pmatrix} \frac{1}{2} & \frac{1}{2} \\ 0 & 1 \end{pmatrix}^{(k-m)/2} = \begin{pmatrix} (\frac{1}{2})^{(k-m)/2} & 1 - (\frac{1}{2})^{(k-m)/2} \\ 0 & 1 \end{pmatrix}.$$

Defining

$$Q_{m,k} = \begin{pmatrix} 0 & 1 \\ 0 & 1 \end{pmatrix},$$

we have

$$\|P^{(m,k)} - Q_{m,k}\| = 2(\tfrac{1}{2})^{(k-m)/2}. \tag{3.2}$$

Similarly, if m is even and k is odd, then

$$P^{(m,k)} = \begin{pmatrix} \frac{1}{2} & \frac{1}{2} \\ 0 & 1 \end{pmatrix}^{[(k-m)/2]} \begin{pmatrix} \frac{1}{2} & \frac{1}{2} \\ 1 & 0 \end{pmatrix}$$

$$= \begin{pmatrix} 1 - (\frac{1}{2})^{(k-m+1)/2} & (\frac{1}{2})^{(k-m+1)/2} \\ 1 & 0 \end{pmatrix}.$$

Defining

$$Q_{m,k} = \begin{pmatrix} 1 & 0 \\ 1 & 0 \end{pmatrix},$$

we have

$$\|P^{(m,k)} - Q_{m,k}\| = 2(\tfrac{1}{2})^{(k-m+1)/2}. \tag{3.3}$$

From (3.2) and (3.3) we see that for m even, $\lim_{k \to \infty} \|P^{(m,k)} - Q_{m,k}\| = 0$. If m is odd a similar argument can be made, hence we may conclude that $\{P_n\}$ is weakly ergodic.

$$P_n = \begin{pmatrix} \tfrac{1}{2} & \tfrac{1}{2} \\ 1 & 0 \end{pmatrix} = \begin{pmatrix} 1 & 0 \\ 1 & 0 \end{pmatrix} + \begin{pmatrix} -\tfrac{1}{2} & \tfrac{1}{2} \\ 0 & 0 \end{pmatrix} = Q_n + R_n$$

and for n even write

$$P_n = \begin{pmatrix} 0 & 1 \\ 1 & 0 \end{pmatrix} = \begin{pmatrix} \tfrac{1}{2} & \tfrac{1}{2} \\ \tfrac{1}{2} & \tfrac{1}{2} \end{pmatrix} + \begin{pmatrix} -\tfrac{1}{2} & \tfrac{1}{2} \\ \tfrac{1}{2} & -\tfrac{1}{2} \end{pmatrix} = Q_n + R_n.$$

Now consider products of the R's and the norms of these products:

$$R_1 = -R_1 R_2, \qquad R_1 R_2 R_3 = \begin{pmatrix} -\tfrac{1}{4} & \tfrac{1}{4} \\ 0 & 0 \end{pmatrix} = -R_1 R_2 R_3 R_4,$$

and

$$R_1 R_2 R_3 R_4 R_5 = \begin{pmatrix} -\tfrac{1}{8} & \tfrac{1}{8} \\ 0 & 0 \end{pmatrix} = -R_1 R_2 R_3 R_4 R_5 R_6,$$

with $\|R_1\| = 1$, $\|R_1 R_2 R_3\| = \tfrac{1}{2}$, and $\|R_1 R_2 R_3 R_4 R_5\| = \tfrac{1}{4}$. In general, $\|\Pi_{n=m}^{m+k} R_n\| < 2(\tfrac{1}{2})^{[(k-1)/2]}$, which tends to zero as $k \to \infty$, for all m. Therefore we may conclude that the chain is weakly ergodic.

We conclude this section with a brief discussion of the stationary case. In this case, the transition matrix at each time is the same so $P^{(m,k)} = P^{k-m}$. By Lemma V.2.3 we have that

$$\delta(P^{(m,k)}) = \delta(P^{k-m}) \leqslant [\delta(P)]^{k-m}.$$

In view of this inequality, a sufficient condition for the stationary chain to be weakly ergodic is that $\delta(P) < 1$. It is not surprising that the condition is not necessary, as is illustrated by the following example.

Example V.3.3. Let $\{X_n\}$ be a stationary chain with transition matrix

$$P = \begin{bmatrix} 0 & 1 & 0 \\ \frac{1}{2} & 0 & \frac{1}{2} \\ \frac{1}{3} & \frac{1}{3} & \frac{1}{3} \end{bmatrix}.$$

It is not hard to see that P is the transition matrix for an irreducible, aperiodic, positive persistent finite Markov chain. It follows that P is weakly ergodic, even though $\delta(P) = 1$.

Since weak ergodicity for a stationary Markov chain is determined by the behavior of the corresponding transition matrix, we will use the phrase weakly ergodic to refer to the chain itself or the corresponding transition matrix. Hence when we say that P is a weakly ergodic stochastic matrix, we mean that the Markov chain determined by P is weakly ergodic.

In Exercise 6, the reader is asked to show that a stationary chain is weakly ergodic if and only if there is some n such that $\delta(P^n) < 1$ [or equivalently $\alpha(P^n) > 0$]. This characterization of weak ergodicity for stationary chains will be used in Chapter VI.

At this point a comparison between the ergodicity described in Chapter III and the weak ergodicity described above is given. In particular, both of these concepts are meaningful for a stationary Markov chain. For finite chains weak ergodicity is equivalent to ergodicity (Exercise 7). However, the following example shows that for infinite chains these concepts are different.

Example V.3.4. Let $\{X_n\}$ be a stationary Markov chain with transition matrix

$$P = \begin{bmatrix} \frac{1}{2} & \frac{1}{2} & 0 & 0 & 0 & \cdot & \cdot & \cdot \\ \frac{1}{2} & 0 & \frac{1}{2} & 0 & 0 & \cdot & \cdot & \cdot \\ 0 & \frac{3}{4} & 0 & \frac{1}{4} & 0 & \cdot & \cdot & \cdot \\ 0 & 0 & \frac{7}{8} & 0 & \frac{1}{8} & \cdot & \cdot & \cdot \\ \cdot & \cdot & \cdot & \cdot & \cdot & \cdot & \cdot & \cdot \\ \cdot & \cdot & \cdot & \cdot & \cdot & \cdot & \cdot & \cdot \\ \cdot & \cdot & \cdot & \cdot & \cdot & \cdot & \cdot & \cdot \end{bmatrix}$$

It can be shown that the chain with transition matrix P, is irreducible, positive persistent, and aperiodic and hence ergodic in the sense of Chapter III. That is $\lim_{n \to \infty} \mathbf{f}_0 P^n = \boldsymbol{\pi}$ independently of $\mathbf{f}_0$. However $\delta(P^m) = 1$ for all m so the chain is not weakly ergodic (Exercise 5).

In view of Example V.3.4 we will be very careful when discussing ergodicity of stationary Markov chains with countably infinite state space. That is, we will use the term weak ergodicity when we have $\delta(P^m) < 1$ for some m. This is definitely a stronger condition than the ergodic behavior discussed in Chapter III.

EXERCISES

1. Let $\{X_n\}$ be a nonstationary Markov chain with transition matrices defined by

$$P_n = \begin{pmatrix} a_n & 1 - a_n \\ 1 - a_n & a_n \end{pmatrix}.$$

Determine whether or not the chain is weakly ergodic when

(a) $a_n = \dfrac{1}{n}$.

(b) $a_n = \dfrac{1}{n^2}$.

2. Let $\{X_n\}$ be a nonstationary Markov chain with transition matrices defined by

$$P_n = \begin{bmatrix} 0 & 1 & 0 \\ 0 & 0 & 1 \\ \frac{3}{4} & \frac{1}{4} & 0 \end{bmatrix} \text{ if } n \text{ is odd,} \qquad P_n = \begin{bmatrix} 0 & 0 & 1 \\ 0 & 0 & 1 \\ \frac{1}{2} & \frac{1}{2} & 0 \end{bmatrix} \text{ if } n \text{ is even.}$$

Determine whether or not the chain is weakly ergodic.

3. Let P and Q be stochastic matrices. Show that if (PQ) is weakly ergodic, then (QP) is weakly ergodic.

4. Show that a weakly ergodic stationary Markov chain has at most one closed set of states and these states are aperiodic.

*5. Show that the stationary Markov chain described in Example V.3.4 is ergodic but not weakly ergodic.

6. Show that a stationary Markov chain with transition matrix P is weakly ergodic if and only if $\delta(P^m) < 1$ for some m.

7. Show that a finite stationary chain with transition matrix P is ergodic if and only if $\delta(P^m) < 1$ for some m.

†**8.**   Let $\{X_n\}$ be a Markov chain with transition matrix

$$
P = \begin{bmatrix}
1 & 0 & 0 & 0 & 0 & 0 & \cdot & \cdot & \cdot \\
1 & 0 & 0 & 0 & 0 & 0 & \cdot & \cdot & \cdot \\
0 & 1 & 0 & 0 & 0 & 0 & \cdot & \cdot & \cdot \\
0 & 0 & 1 & 0 & 0 & 0 & \cdot & \cdot & \cdot \\
\cdot & \cdot & \cdot & \cdot & \cdot & \cdot & \cdot & \cdot & \cdot \\
\cdot & \cdot & \cdot & \cdot & \cdot & \cdot & \cdot & \cdot & \cdot \\
\cdot & \cdot & \cdot & \cdot & \cdot & \cdot & \cdot & \cdot & \cdot
\end{bmatrix}
$$

(a) Show that P is ergodic.
(b) Show that P is not weakly ergodic.
(c) Find the mean time until absorption from state j. Call this m_{j1}.
(d) Let $\nu = (\nu_1, \nu_2, \nu_3, \ldots)$ be a starting vector and define $m_{\nu 1}$ $= \sum_{j=1}^{\infty} \nu_j m_{j1}$. Find a starting vector ν such that $m_{\nu 1} = \infty$.
(e) Show that a stochastic matrix, P, is weakly ergodic if and only if there exists a state j such that j is aperiodic and $m_{\nu j} < \infty$ for all starting vectors ν.

SECTION 4   **STRONG ERGODICITY**

As in the case of weak ergodicity, it is usually difficult to show that a nonstationary Markov chain is strongly ergodic directly from the definition. In this section we present some theorems that give sufficient conditions for a chain to be strongly ergodic. These theorems also may be easier to use than the definition.

The first result we give is actually a simple variation of Definition V.1.4. It relates strongly ergodic behavior to the transition matrices rather than the starting vectors.

Theorem V.4.1.

A nonstationary Markov chain with transition matrices $\{P_n\}$ is strongly ergodic if and only if there exists a constant matrix Q such that for each m

$$
\lim_{k \to \infty} \| P^{(m,k)} - Q \| = 0.
$$

Proof. The proof is straightforward if one uses the first row of Q as the vector $\mathbf{q}$, and conversely (Exercise 13). ▲

The next theorem we consider is due to Paz (1970).

Theorem V.4.2.

A nonstationary Markov chain is strongly ergodic if and only if there is a sequence of constant stochastic matrices $\{Q_m\}$ and, for each m, there is a sequence of constant stochastic matrices $\{Q_{mk}\}$ such that

$$\text{(i)} \quad \lim_{k \to \infty} \|P^{(m,k)} - Q_{mk}\| = 0 \quad \text{and}$$

$$\text{(ii)} \quad \lim_{k \to \infty} \|Q_{mk} - Q_m\| = 0.$$

Proof. Assume sequences of constant matrices $\{Q_m\}$ and $\{Q_{mk}\}$ satisfying conditions (i) and (ii) exist. Since

$$\|P^{(m,k)} - Q_m\| \leqslant \|P^{(m,k)} - Q_{mk}\| + \|Q_{mk} - Q_m\|,$$

it follows that for all m, $\lim_{k \to \infty} \|P^{(m,k)} - Q_m\| = 0$. Clearly, if $Q_m = Q$ for all m, then by Theorem V.4.1., the chain will be strongly ergodic. In other words, it suffices to show that Q_m is the same constant matrix for all m. It is easy to show that $P_m Q_m = Q_m$ (Exercise 11). We also know from Lemma V.2.5 that for any two matrices, A and B, $\|AB\| \leqslant \|A\| \cdot \|B\|$. Hence we get

$$\|Q_{m-1} - Q_m\| \leqslant \|Q_{m-1} - P^{(m-1,k)}\| + \|P_m \cdot P^{(m,k)} - P_m Q_m\| + \|P_m Q_m - Q_m\|$$

$$= \|Q_{m-1} - P^{(m-1,k)}\| + \|P_m(P^{(m,k)} - Q_m)\|$$

$$\leqslant \|Q_{m-1} - P^{(m-1,k)}\| + \|P_m\| \cdot \|P^{(m,k)} - Q_m\|$$

$$\leqslant \|Q_{m-1} - P^{(m-1,k)}\| + \|P^{(m,k)} - Q_m\|.$$

By letting $k \to \infty$, we get that $\|Q_{m-1} - Q_m\| = 0$, which implies that $Q_{m-1} = Q_m$ for all m.

Conversely, if the chain is strongly ergodic, then by setting $Q_m = Q_{mk} = Q$ for all m and k, it follows that (i) and (ii) are true. ▲

Theorems V.4.1 and V.4.2 both have an advantage over the original definition of strong ergodicity in that the supremum over all possible starting vectors need not be considered. However, the advantage is somewhat illusory since finding the norm of a matrix involves consideration of all of the rows.

In actually trying to determine whether a nonstationary Markov chain is strongly ergodic, it seems that Theorem V.4.1 would be most useful. However, in order to try to use this theorem, we must have a candidate for the constant matrix Q. Once the candidate is presented, we must of course see whether or not $\|P^{(m,k)} - Q\| \to 0$ as $k \to \infty$ to see if it was the "correct" candidate. How do we go about finding a candidate? We can get some indication by considering the stationary case.

If a stationary Markov chain is ergodic, the powers of the transition matrix, P^n, converge to a constant matrix, Q. The rows of Q are all equal to the invariant probability vector associated with P. That is, if the vector ψ represents a row of Q, then $\psi P = \psi$ and $\|\psi\| = 1$. Hence the rows of Q are the unique left eigenvector of P corresponding to the eigenvalue $\lambda = 1$. In view of the relationship that exists between left eigenvectors of P and the constant matrix Q in the stationary case, we might suspect that some relationship exists between the left eigenvectors ψ_n of P_n and Q in the nonstationary case. We will explore that relationship in the remainder of this section.

It follows from Exercise 4 of Section III.5 that there is a class of infinite stochastic matrices for which there is no left eigenvector of the eigenvalue 1 satisfying $\psi_i \geqslant 0$ for all i and $\|\psi\| = 1$. In view of this, we will restrict attention to those matrices that do have such eigenvectors.

Definition V.4.1. *Let $\mathcal{Q}$ be the class of stochastic matrices P for which there exists at least one non-negative left eigenvector ψ, corresponding to the eigenvalue 1, such that $\|\psi\| = 1$.*

If $P \in \mathcal{Q}$, then whenever we refer to a left eigenvector of P, we will be referring to an eigenvector ψ satisfying the above conditions. Note that such an eigenvector always exists for finite stochastic matrices (Exercise 18).

One might ask at this point whether $\mathcal{Q}$ should be restricted further to those stochastic matrices having a *unique* eigenvector satisfying the above conditions. This would be equivalent to requiring that the matrices have exactly one closed subset of persistent states (actually, that they correspond to stationary Markov chains which have this property), and this would be an unnecessary restriction. In particular, such a restriction would disallow the identity matrix from appearing in the sequence of matrices $\{P_n\}$ of a nonstationary Markov chain. (Why?) However, since there may be certain steps in a nonstationary chain at which no change takes place, it would be appropriate to allow the identity to be a member of the sequence $\{P_n\}$. The reader should keep in mind this possible non-uniqueness of the left eigenvectors while reading the following theorems.

Let $\{P_n\}_{n=1}^{\infty}$ be a sequence of transition matrices corresponding to a nonstationary Markov chain. If $P_n \in \mathcal{Q}$ for all n, then there is at least one associated sequence of corresponding left eigenvectors, ψ_n. Consideration of the stationary case makes it reasonable to ask whether strong ergodicity of the chain is related to convergence of the sequence of left eigenvectors,

ψ_n. If in fact ψ_n converged to a vector ψ, then the constant matrix Q, having all of its rows equal to ψ, would be a candidate for the constant matrix needed in Theorem V.4.1. Theorems V.4.3 and V.4.4 give conditions under which this candidate is the correct choice.

We know (from Exercise 5 of Section V.2) that weak ergodicity is necessary for strong ergodicity. In the next theorems, we assume sufficient conditions for weak ergodicity and then impose some additional conditions on $\{\psi_n\}$ to obtain sufficient conditions for strong ergodicity.

Theorem V.4.3.

Let $\{P_n\}$ be a sequence of transition matrices corresponding to a non-stationary weakly ergodic Markov chain with $P_n \in \mathcal{Q}$ for all n. If there exists a corresponding sequence of left eigenvectors ψ_n satisfying

$$\sum_{j=1}^{\infty} \|\psi_j - \psi_{j+1}\| < \infty, \tag{4.1}$$

then the chain is strongly ergodic.

Proof. The condition imposed on the left eigenvectors is stronger than assuming $\{\psi_n\}_{n=1}^{\infty}$ converges in norm to some vector, ψ (Exercise 19). Hence we can define $\psi = \lim_{n \to \infty} \psi_n$ and note that $\|\psi_n - \psi\| \to 0$ as $n \to \infty$. Since all of the ψ_n's have the property that their components are non-negative and add to one, ψ will also have this property. Define Q to be the constant stochastic matrix with each row equal to ψ. In order to show $\{P_n\}$ is strongly ergodic it is sufficient to show $\|P^{(m,k)} - Q\| \to 0$ as $k \to \infty$ for all m. For notational convenience let Q_n denote the constant stochastic matrix with rows equal to ψ_n. Let m be fixed. Using the triangle inequality and the fact that $P^{(m,k)} = P^{(m,l)}P^{(l,k)}$ we get

$$\|P^{(m,k)} - Q\| \leqslant \|P^{(m,k)} - Q_k\| + \|Q_k - Q\|$$

$$\leqslant \|P^{(m,l)}P^{(l,k)} - Q_{l+1}P^{(l,k)}\| + \|Q_{l+1}P^{(l,k)} - Q_k\| + \|Q_k - Q\|. \tag{4.2}$$

In order to prove that $\lim_{k \to \infty} \|P^{(m,k)} - Q\| = 0$ we let $\epsilon > 0$ be given and show that there exists K such that for all $k \geqslant K$, $\|P^{(m,k)} - Q\| < \epsilon$. We do this by making each of the three terms on the right-hand side of (4.2) less than $\epsilon/3$.

We first consider the middle term of the right-hand side of (4.2) and note that since $Q_{l+1}P_{l+1} = Q_{l+1}$ we have

$$Q_{l+1}P^{(l,k)} = Q_{l+1}P^{(l+1,k)}$$

$$= Q_{l+1}P^{(l+1,k)} - Q_{l+2}P^{(l+1,k)} + Q_{l+2}P^{(l+1,k)}$$

$$= (Q_{l+1} - Q_{l+2})P^{(l+1,k)} + Q_{l+2}P^{(l+1,k)}.$$

Repeating this procedure on $Q_{l+2}P^{(l+1,k)}$, we get

$$Q_{l+1}P^{(l,k)} = (Q_{l+1} - Q_{l+2})P^{(l+1,k)} + (Q_{l+2} - Q_{l+3})P^{(l+2,k)} + Q_{l+3}P^{(l+2,k)}.$$

If we continue this process until the last term is of the form $Q_k P_k$, and note that $Q_k P_k = Q_k$, we will have

$$Q_{l+1}P^{(l,k)} = \sum_{j=l+1}^{k-1} (Q_j - Q_{j+1})P^{(j,k)} + Q_k.$$

Hence using the triangle inequality, Lemma V.2.4, and the fact that $\delta(P^{(j,k)}) \leqslant 1$ we get

$$\|Q_{l+1}P^{(l,k)} - Q_k\| = \left\| \sum_{j=l+1}^{k-1} (Q_j - Q_{j+1})P^{(j,k)} \right\|$$

$$\leqslant \sum_{j=l+1}^{k-1} \|(Q_j - Q_{j+1})P^{(j,k)}\|$$

$$\leqslant \sum_{j=l+1}^{k-1} \|Q_j - Q_{j+1}\| \delta(P^{(j,k)})$$

$$\leqslant \sum_{j=l+1}^{k-1} \|Q_j - Q_{j+1}\|. \tag{4.3}$$

Since by construction Q_j has all of its rows equal to ψ_j, it follows that $\|Q_j - Q_{j+1}\| = \|\psi_j - \psi_{j+1}\|$. Hence, using assumption (4.1), we can choose $l^* > m$ such that for all $k > l^*$,

$$\|Q_{l^*+1}P^{(l^*,k)} - Q_k\| \leqslant \sum_{j=l^*+1}^{k-1} \|Q_j - Q_{j+1}\|$$

$$= \sum_{j=l^*+1}^{k-1} \|\psi_j - \psi_{j+1}\| < \frac{\epsilon}{3}.$$

With l^* fixed, next consider the first term of the right-hand side of (4.2). Since $P^{(m,l^*)}$ and Q_{l^*+1} are stochastic matrices, it follows that $\|P^{(m,l^*)} - Q_{l^*+1}\| \leqslant 2$, so by Lemma V.2.4

$$\|P^{(m,l^*)}P^{(l^*,k)} - Q_{l^*+1}P^{(l^*,k)}\| \leqslant \|P^{(m,l^*)} - Q_{l^*+1}\|\delta(P^{(l^*,k)})$$

$$\leqslant 2\delta(P^{(l^*,k)}).$$

Using the assumption that the chain is weakly ergodic, we can find $K_1 > l^*$, such that for all $k \geqslant K_1$, $\delta(P^{(l^*,k)}) < \epsilon/6$. For such values of k,

$$\|P^{(m,l^*)}P^{(l^*,k)} - Q_{l^*+1}P^{(l^*,k)}\| < \frac{\epsilon}{3}.$$

For the third term on the right-hand side of (4.2) we simply note that ψ_k converges in norm to ψ by Exercise 19 and so $\lim_{k\to\infty}\|Q_k - Q\| = 0$. Hence there exists K_2 such that for all $k \geqslant K_2$ we have $\|Q_k - Q\| < \epsilon/3$. Therefore for $k \geqslant \max(K_1, K_2)$ we have

$$\|P^{(m,k)} - Q\| < \frac{\epsilon}{3} + \frac{\epsilon}{3} + \frac{\epsilon}{3} = \epsilon. \qquad \blacktriangle$$

***Remark* V.4.1.** Since the theorem only requires the existence of a sequence of corresponding left eigenvectors with $\sum_{j=1}^{\infty}\|\psi_j - \psi_{j+1}\| < \infty$, the inclusion of some identity matrices in $\{P_n\}$ causes no problems. That is, all vectors are left eigenvectors of I so if $P_j = I$ simply take $\psi_j = \psi_{j+1}$ so the jth step adds nothing to $\sum_{j=1}^{\infty}\|\psi_j - \psi_{j+1}\|$.

Recall that one goal of this section is to relate strong ergodicity to convergence of left eigenvectors. Unfortunately, the condition placed on the left eigenvectors in Theorem V.4.3 is much stronger than assuming convergence of ψ_n to ψ. The next theorem does relax this condition on the left eigenvectors, but the assumption of weak ergodicity is replaced by a stronger condition.

Theorem V.4.4.

Let $\{P_n\}$ be a sequence of transition matrices corresponding to a nonstationary Markov chain with $P_n \in \mathcal{C}$ for all n. If there exists a corresponding sequence of left eigenvectors $\{\psi_n\}$ and a vector ψ satisfying $\|\psi_n - \psi\| \to 0$ as $n \to \infty$, and if there exists a constant $D < \infty$ such that

$$\sum_{j=0}^{k-1} \delta(P^{(j,k)}) \leqslant D \qquad (4.4)$$

for all k, then the chain is strongly ergodic.

Proof. It can be shown that if inequality (4.4) holds for all k, then $\delta(P^{(m,k)}) \to 0$ as $k \to \infty$, but of course the converse is not true, so the condition required in this theorem is stronger than weak ergodicity (Exercise 14).

Let Q and Q_n be as defined in Theorem V.4.3. We will again show that for each m, $\|P^{(m,k)} - Q\| \to 0$ as $k \to \infty$. Let m be fixed and let $\epsilon > 0$ be given. By using the arguments given in the proof of Theorem V.4.3, it suffices to show that each of the three terms on the right-hand side of (4.2) is less than $\epsilon/3$. In order to show that the middle term is less than $\epsilon/3$, we consider inequality (4.3) and note that

$$\|Q_{l+1}P^{(l,k)} - Q_k\| \leqslant \sum_{j=l+1}^{k-1} \|Q_j - Q_{j+1}\|\delta(P^{(j,k)}).$$

At this point the proof differs from the proof of Theorem V.4.3. Since we now assume $\|\psi_n - \psi\| \to 0$ as $n \to \infty$, it follows from the definition of Q and Q_n that there exists an l^* such that for all $j \geqslant l^* + 1$ we have $\|Q_j - Q_{j+1}\| < \epsilon/3D$. For this choice of l^* and for all $k > l^* + 1$, we have

$$\sum_{j=l^*+1}^{k-1} \|Q_j - Q_{j+1}\|\delta(P^{(j,k)}) < \frac{\epsilon}{3D} \sum_{j=l^*+1}^{k-1} \delta(P^{(j,k)}) < \frac{\epsilon}{3}. \qquad (4.5)$$

We conclude the proof by noting that for the fixed l^*, the first and third terms of the right-hand side of (4.2) can be shown to be less than $\epsilon/3$ by the same arguments as given in the proof of Theorem V.4.3. ▲

The condition on the left eigenvectors in Theorem V.4.3 and the condition on the delta coefficient in Theorem V.4.4 are strong conditions. The fact that they seem somewhat artificial leads one naturally to the following conjecture. If $\{P_n\}$ in $\mathcal{C}$ is weakly ergodic and if there exists a corresponding sequence of left eigenvectors $\{\psi_n\}$ and a vector ψ with $\|\psi_n - \psi\| \to 0$ as $n \to \infty$, then $\{P_n\}$ is strongly ergodic. Unfortunately this conjecture is not true as the next example shows.

Example V.4.1. Let $\{X_n\}$ be a nonstationary Markov chain with transition matrices defined to be

$$P_{2n-1} = \begin{pmatrix} 0 & 1 \\ 1 & 0 \end{pmatrix} \quad \text{and} \quad P_{2n} = \begin{pmatrix} 0 & 1 \\ 1 - \dfrac{1}{2n} & \dfrac{1}{2n} \end{pmatrix} \quad \text{for } n = 1, 2, 3, \dots.$$

The corresponding left eigenvectors are

$$\psi_{2n-1} = (\tfrac{1}{2}, \tfrac{1}{2}) \quad \text{and} \quad \psi_{2n} = \left(\frac{2n-1}{4n-1}, \frac{2n}{4n-1} \right).$$

It is easy to show that this chain is weakly ergodic and $\lim_{n\to\infty}\psi_n=(\frac{1}{2},\frac{1}{2})$ (Exercise 2). Consider the product $P_1P_2\cdots P_{2n-1}P_{2n}$ and express the product using pairs $(P_{2k-1}P_{2k})$. It follows that

$$P_{2k-1}P_{2k}=\begin{bmatrix} 1-\dfrac{1}{2k} & \dfrac{1}{2k} \\ 0 & 1 \end{bmatrix} \quad \text{for } k=1,2,3,\ldots.$$

Now the left eigenvector corresponding to $P_{2k-1}P_{2k}$ is $(0,1)$ for $k=1,2,3,\ldots$. If we define a new chain with transition matrices $R_k=P_{2k-1}P_{2k}$, it follows from Theorem V.4.3 that $\{R_k\}$ is strongly ergodic with

$$R_1R_2\cdots R_k\to\begin{pmatrix} 0 & 1 \\ 0 & 1 \end{pmatrix} \quad \text{as } k\to\infty.$$

That is

$$(P_1P_2)(P_3P_4)\cdots(P_{2k-1}P_{2k})\to\begin{pmatrix} 0 & 1 \\ 0 & 1 \end{pmatrix} \quad \text{as } k\to\infty.$$

On the other hand, if we define $S_k=P_{2k}P_{2k+1}$ we see that the left eigenvector corresponding to S_k is $(1,0)$ for $k=1,2,3,\ldots$. It again follows from Theorem V.4.3 that $\{S_k\}$ is strongly ergodic and

$$P_1(P_2P_3)(P_4P_5)\cdots(P_{2n}P_{2n+1})\to\begin{pmatrix} 1 & 0 \\ 1 & 0 \end{pmatrix} \quad \text{as } n\to\infty.$$

Hence $\{P_n\}$ is not strongly ergodic (why?) even though the hypotheses of the conjecture are satisfied.

In view of this example the extra conditions of Theorems V.4.3 and V.4.4 are easier to accept. However, these conditions have not been shown to be necessary, so there may be yet a better set of sufficient conditions. The question of necessary conditions for strong ergodicity is discussed briefly later.

We consider the implications of Theorems V.4.3 and V.4.4 in the case of stationary Markov chains.

Corollary V.4.1. Let $P\in\mathcal{C}$ be the transition matrix of a stationary Markov chain. If the chain is weakly ergodic, then it is strongly ergodic.

Proof. In the stationary case $P_n=P$ for all n and $\psi_n=\psi$ for all n, so it follows immediately that

$$\sum_{j=1}^{\infty} \|\psi_j-\psi_{j+1}\|=0<\infty.$$

Therefore it follows from Theorem V.4.3 that the chain is strongly ergodic.
▲

Corollary V.4.1 can also be proved by using Theorem V.4.4. This is left as an exercise for the reader.

In view of Corollary V.4.1 it is now clear why the concepts of weak and strong ergodicity are not separated in the case of stationary Markov chains. It is impossible for a stationary chain to be weakly ergodic without being strongly ergodic; that is, it is impossible for a stationary chain to lose memory without converging. However, we note that it is possible to have convergence without losing memory. As a simple example, if the transition matrix for a stationary chain is given by $P = \begin{pmatrix} 1 & 0 \\ 0 & 1 \end{pmatrix}$, then $\mathbf{f}^{(k)} = \mathbf{f}^{(0)}$ for all k, so $\mathbf{f}^{(k)}$ trivially converges, but memory is not lost.

***Remark* V.4.2.** In Theorem VI.4.2 we show that a weakly ergodic stationary Markov chain must have a unique left invariant probability vector. Therefore the assumption in Corollary V.4.1 that $P \in \mathcal{C}$ is redundant.

We should point out that Hajnal (1956) and Mott (1957) have proved theorems similar to Theorems V.4.3 and V.4.4. However, they restricted their attention to finite Markov chains and did not use the δ coefficient as developed by Dobrushin or the norm of a matrix used in this chapter. By using the δ coefficient and the norm of a matrix we have been able to give results that are applicable to countably infinite Markov chains. In fact, Theorems V.4.3 and V.4.4 have been proven for discrete-time Markov processes where the state space is quite arbitrary (Madsen and Isaacson, 1973). In such cases the transition matrices are replaced by transition functions. However, an understanding of measure theory is required when the state space is not discrete, so the general results will not be given here.

We continue this section with some examples in which Theorems V.4.3 and V.4.4 can be used to show strong ergodicity.

Example V.4.2. Let $\{X_n\}$ be a nonstationary Markov chain with transition matrices given by

$$P_{2k-1} = \begin{bmatrix} \frac{1}{2} & 0 & \frac{1}{2} \\ \frac{1}{2} & \frac{1}{2} & 0 \\ 0 & \frac{1}{2} & \frac{1}{2} \end{bmatrix}, \qquad P_{2k} = \begin{bmatrix} \frac{1}{3} & \frac{1}{3} & \frac{1}{3} \\ \frac{1}{6} & \frac{1}{3} & \frac{1}{2} \\ \frac{1}{2} & \frac{1}{3} & \frac{1}{6} \end{bmatrix}$$

for $k = 1, 2, 3, \ldots$. It is easy to see that the chain is weakly ergodic since $\delta(P_{2k}) = \frac{1}{3}$ and $\delta(P_{2k-1}) = \frac{1}{2}$. We can see that $\psi_n = (\frac{1}{3}, \frac{1}{3}, \frac{1}{3})$ is a left eigenvector for all of these matrices, so by Theorem V.4.3 it follows that the chain is strongly ergodic.

Example V.4.3. A simple generalization of Example V.4.2 can be made as follows. Let $\{X_n\}$ be a nonstationary Markov chain with state space $S = \{1, 2, \ldots, N\}$. Let the corresponding transition matrices be given by $\{P_n\}$. If all of the matrices of this sequence are doubly stochastic (that is, if $\sum_{j=1}^{N} P_{ij} = 1$ and $\sum_{i=1}^{N} P_{ij} = 1$), then the chain is strongly ergodic if and only if it is weakly ergodic. To see this, note that $\psi_n = (1/N, 1/N, \ldots, 1/N)$ is a left eigenvector for each P_n, so that $\sum_{n=1}^{\infty} \|\psi_n - \psi_{n+1}\| < \infty$. It follows from Theorem V.4.3 that if the chain is weakly ergodic, it will also be strongly ergodic.

It is certainly not true that all sequences of doubly stochastic matrices are weakly ergodic. This is easy to see since the identity matrix is doubly stochastic, but a sequence of identities would not be weakly ergodic.

Example V.4.4. Let $\{X_n\}$ be a nonstationary Markov chain with transition matrices

$$P_1 = \begin{pmatrix} \frac{1}{3} & \frac{2}{3} \\ \frac{1}{2} & \frac{1}{2} \end{pmatrix}, \qquad P_n = \begin{pmatrix} \dfrac{1}{3} + \dfrac{1}{n} & \dfrac{2}{3} - \dfrac{1}{n} \\ \dfrac{1}{2} & \dfrac{1}{2} \end{pmatrix}$$

for $n \geq 2$. Since $\delta(P_n) = \frac{1}{6} - 1/n$ for $n \geq 6$, it follows that the chain is weakly ergodic. The left eigenvector corresponding to P_n ($n \geq 2$) is

$$\psi_n = \frac{3n}{7n-6}\left(1, \frac{4n-6}{3n}\right) = \left(\frac{3n}{7n-6}, \frac{4n-6}{7n-6}\right).$$

Consequently,

$$\|\psi_n - \psi_{n+1}\| = \left| \frac{3n}{7n-6} - \frac{3(n+1)}{7(n+1)-6} \right| + \left| \frac{4n-6}{7n-6} - \frac{4(n+1)-6}{7(n+1)-6} \right|$$

$$= \left| \frac{18}{(7n-6)(7n+1)} \right| + \left| \frac{-6}{(7n-6)(7n+1)} \right| \leq \frac{1}{n^2}$$

for all $n \geq 2$. Since $\sum_{n=2}^{\infty} 1/n^2 < \infty$, it follows from Theorem V.4.3 that the chain is strongly ergodic.

The chain given in this example can also be shown to be strongly ergodic using Theorem V.4.4 (Exercise 5).

Example V.4.5. Let $\{X_n\}$ be a nonstationary Markov chain with transition matrices given by

$$P_{2n-1} = \begin{pmatrix} \frac{1}{2} & \frac{1}{2} \\ 1 & 0 \end{pmatrix}, \qquad P_{2n} = \begin{pmatrix} \dfrac{1}{2} + \dfrac{1}{2n} & \dfrac{1}{2} - \dfrac{1}{2n} \\ 1 & 0 \end{pmatrix}$$

for $n = 1, 2, 3, \ldots$. The left eigenvector corresponding to P_{2n} is

$$\psi_{2n} = \left(\frac{2n}{3n-1}, \frac{n-1}{3n-1} \right)$$

and the left eigenvector corresponding to P_{2n-1} is $\psi_{2n-1} = (\frac{2}{3}, \frac{1}{3})$. Hence we have $\psi_n \to \psi = (\frac{2}{3}, \frac{1}{3})$. Furthermore, $\delta(P_{2n}) = \frac{1}{2} - 1/(2n)$, and $\delta(P_{2n-1}) = \frac{1}{2}$, so $\delta(P_n) \leqslant \frac{1}{2}$ for all n.

By Lemma V.2.3 we have that

$$\delta(P^{(j,k)}) = \delta(P_{j+1} \cdot P_{j+2} \cdot \cdots \cdot P_k) \leqslant \prod_{i=j+1}^{k} \delta(P_i) \leqslant (\tfrac{1}{2})^{k-j}.$$

Consequently

$$\sum_{j=0}^{k-1} \delta(P^{(j,k)}) \leqslant \sum_{j=0}^{k-1} (\tfrac{1}{2})^{k-j} = \sum_{i=1}^{k} (\tfrac{1}{2})^{i} \leqslant 1$$

for all k. Hence by Theorem V.4.4 we may conclude that the chain is strongly ergodic.

Remark V.4.3. The reader will be asked to show that the hypotheses of Theorem V.4.3 are not satisfied for the Markov chain defined in Example V.4.5 (Exercise 4), so Theorem V.4.3 can not be used to show strong ergodicity. By using a special case of Example V.4.3, we can exhibit a strongly ergodic nonstationary Markov chain for which the hypotheses of Theorem V.4.3 are satisfied while the hypotheses of Theorem V.4.4 are not satisfied (Exercise 3). Hence we can conclude that neither of these two theorems implies the other.

The reader will recall that the original motivation in giving Theorems V.4.3 and V.4.4 was to provide a means for determining whether a nonstationary chain is strongly ergodic. We would hope that it would be easier to check the conditions of these theorems than it would be to use the definition of strong ergodicity itself. Unfortunately, the conditions of Theorem V.4.3 and V.4.4 require knowledge of the left eigenvectors, ψ_n, of each P_n. Since the calculation of the eigenvectors is generally a nontrivial problem, there may not be much gained by using these theorems instead of the definition. In view of this, we will give a set of sufficient conditions for strong ergodicity that does not require knowledge of the eigenvectors.

We know that if a stochastic matrix P is weakly ergodic then the stationary chain generated by P is strongly ergodic. If we view this chain as a "nonstationary" chain with $P_n = P$ for all n, then we have a trivial example where $\| P_n - P \| \to 0$ as $n \to \infty$ and P is weakly ergodic implies that the nonstationary chain is strongly ergodic. We might wonder if the same result holds when the convergence in norm of P_n to P is nontrivial. Mott (1957) proved that this result is true in the case of finite Markov chains. We will prove that the result is in fact true if the chain is finite or countably infinite. [The result is even true for arbitrary Markov processes, but this case will not be discussed here (Bowerman, David, and Isaacson (1975)).] The proof of this result is based on a series of lemmas that are given below.

Lemma V.4.1.

If P and Q are stochastic matrices, then

$$|\delta(P) - \delta(Q)| \leqslant \|P - Q\|.$$

Proof.

$$|\delta(P) - \delta(Q)| = \left| \frac{1}{2} \sup_{i,k} \sum_{j=1}^{\infty} |p_{ij} - p_{kj}| - \frac{1}{2} \sup_{i,k} \sum_{j=1}^{\infty} |q_{ij} - q_{kj}| \right|$$

$$\leqslant \left| \frac{1}{2} \sup_{i,k} \left\{ \sum_{j=1}^{\infty} |p_{ij} - p_{kj}| - |q_{ij} - q_{kj}| \right\} \right|$$

$$\leqslant \frac{1}{2} \sup_{i,k} \sum_{j=1}^{\infty} |(p_{ij} - p_{kj}) - (q_{ij} - q_{kj})|$$

$$= \frac{1}{2} \sup_{i,k} \sum_{j=1}^{\infty} |(p_{ij} - q_{ij}) - (p_{kj} - q_{kj})|$$

$$\leqslant \frac{1}{2} \sup_{i,k} \sum_{j=1}^{\infty} (|p_{ij} - q_{ij}| + |p_{kj} - q_{kj}|)$$

$$= \frac{1}{2} \sup_{i} \sum_{j=1}^{\infty} |p_{ij} - q_{ij}| + \frac{1}{2} \sup_{k} \sum_{j=1}^{\infty} |p_{kj} - q_{kj}|$$

$$= \|P - Q\|. \quad \blacktriangle$$

Lemma V.4.2. If $\|P_n - P\| \to 0$ as $n \to \infty$, then $\delta(P_n) \to \delta(P)$ as $n \to \infty$.

Proof. It follows from Lemma V.4.1 that $|\delta(P_n) - \delta(P)| \leqslant \|P_n - P\| \to 0$, as $n \to \infty$. $\quad \blacktriangle$

Lemma V.4.3. If $\{P_n\}$ is a sequence of stochastic matrices for which $\|P_n - P\| \to 0$ as $n \to \infty$, then for any positive integer k, $\|P_n^k - P^k\| \to 0$ as $n \to \infty$.

Proof. The lemma can be proved by induction. We will do the case where $k = 2$ and leave the remainder of the proof as an exercise. Using the triangle inequality for norms and Lemma V.2.5 we have that

$$\|P_n^2 - P^2\| \leqslant \|P_n^2 - P_n P\| + \|P_n P - P^2\|$$

$$= \|P_n(P_n - P)\| + \|(P_n - P)P\|$$

$$\leqslant \|P_n\| \cdot \|P_n - P\| + \|P\| \cdot \|P_n - P\|$$

$$= 2\|P_n - P\|.$$

The last equality is true since the norm of a stochastic matrix is one. Clearly the last term goes to zero as $n \to \infty$ by the hypothesis. ▲

Lemma V.4.4. If $\{P_n\}$ is a sequence of stochastic matrices in $\mathcal{R}$ such that $\|P_n - P\| \to 0$ as $n \to \infty$, and if P is weakly ergodic, then there is a sequence of left eigenvectors ψ_n of P_n, and a left eigenvector ψ, of P such that $\|\psi_n - \psi\| \to 0$ as $n \to \infty$.

Proof. For notational convenience let Q_n and Q denote constant matrices with rows ψ_n and ψ, respectively. [A weakly ergodic stochastic matrix has a *unique* left eigenvector, so ψ is unique (see Remark V.4.2).] It follows that

$$\|\psi_n - \psi\| = \|Q_n - Q\| = \|Q_n P_n^k - Q P^k\|$$

for all positive integers k. Hence

$$\|\psi_n - \psi\| \leqslant \|Q_n P_n^k - Q_n P^k\| + \|Q_n P^k - Q P^k\|$$

$$= \|Q_n(P_n^k - P^k)\| + \|(Q_n - Q)P^k\|. \qquad (4.6)$$

Let $\epsilon > 0$ be given. Since P is weakly ergodic there exists k^* such that $\delta(P^{k^*}) < \epsilon/4$. Note that for such a k^*, it follows from Lemma V.2.4 that

$$\|(Q_n - Q)P^{k^*}\| \leqslant \|Q_n - Q\|\delta(P^{k^*}) < \frac{\epsilon}{2}.$$

Next, with k^* fixed, we can conclude from Lemma V.4.3 that $\|P_n^{k^*} - P^{k^*}\| \to 0$ as $n \to \infty$. Hence there exists N such that for all $n \geqslant N$, $\|P_n^{k^*} - P^{k^*}\| < \epsilon/2$. From Lemma V.2.5 we conclude that $\|Q_n(P_n^{k^*} - P^{k^*})\| < \epsilon/2$ for

$n \geq N$, so it follows that the right-hand side of (4.6) is less than ϵ for all $n > N$. ▲

The following example shows that not every convergent sequence of stochastic matrices has converging left eigenvectors. That is, if $\|P_n - P\| \to 0$ as $n \to \infty$ and if P is *not* weakly ergodic then the associated eigenvectors may fail to converge.

Example V.4.6. Define

$$
P_{2n} = \begin{bmatrix} 1 - \dfrac{1}{2n} & \dfrac{1}{2n} \\ \dfrac{1}{2n} & 1 - \dfrac{1}{2n} \end{bmatrix} \quad \text{and} \quad P_{2n-1} = \begin{bmatrix} 1 - \dfrac{2}{2n+1} & \dfrac{2}{2n+1} \\ \dfrac{1}{2n+1} & 1 - \dfrac{1}{2n+1} \end{bmatrix}
$$

for $n = 1, 2, \ldots$. It is not hard to see that $\psi_{2n} = (\tfrac{1}{2}, \tfrac{1}{2})$ and $\psi_{2n-1} = (\tfrac{1}{3}, \tfrac{2}{3})$ for $n = 1, 2, \ldots$, so the left eigenvectors do not converge even though

$$
\lim_{n \to \infty} P_{2n} = \lim_{n \to \infty} P_{2n-1} = \begin{pmatrix} 1 & 0 \\ 0 & 1 \end{pmatrix}.
$$

Lemma V.4.5. If $\{P_n\}$ is a sequence of stochastic matrices such that $\|P_n - P\| \to 0$ as $n \to \infty$, then for each positive integer, k,

$$
\|P_{n+1} \cdot P_{n+2} \cdot \; \cdots \; \cdot P_{n+k} - P^k\| \to 0
$$

as $n \to \infty$.

Proof. The lemma can be proved by induction. The proof is left as an exercise. ▲

We now use Lemmas V.4.1–V.4.5 to prove the following theorem.

Theorem V.4.5.

Let $\{P_n\}$ be a sequence of transition matrices corresponding to a non-stationary Markov chain with $P_n \in \mathcal{C}$. If $\|P_n - P\| \to 0$ as $n \to \infty$ where P is weakly ergodic, then the chain is strongly ergodic.

Proof. We will show that the hypotheses of this theorem imply the hypotheses of Theorem V.4.4 and consequently the chain must be strongly ergodic.

Since the hypotheses of this theorem are the same as the hypotheses of Lemma V.4.4, it follows that there exists a sequence of left eigenvectors $\{\psi_n\}$, corresponding to $\{P_n\}$, and a vector ψ, the eigenvector of P, satisfying $\|\psi_n - \psi\| \to 0$ as $n \to \infty$. Therefore we only need to show that there exists a positive number D such that

$$\sum_{j=0}^{k-1} \delta(P^{(j,k)}) \leqslant D \qquad (4.7)$$

for all k.

Since P is weakly ergodic, there exists a positive integer m such that $\delta(P^m) \leqslant \frac{1}{2}$. It follows from Lemmas V.4.5 and V.4.2 that $\lim_{n\to\infty} \delta(P_{n+1} \cdot P_{n+2} \cdot \cdots \cdot P_{n+m}) \leqslant \frac{1}{2}$. Hence there exists an integer N such that for all $n \geqslant N$,

$$\delta(P_{n+1} \cdot P_{n+2} \cdot \cdots \cdot P_{n+m}) \leqslant \frac{3}{4}. \qquad (4.8)$$

(The reader will note that there is no special significance to the numbers $\frac{1}{2}$ and $\frac{3}{4}$. These numbers are just chosen for explicitness.)

Now consider $\sum_{j=0}^{k-1} \delta(P^{(j,k)})$. If $k \leqslant N + m$, then, since $\delta(P^{(j,k)}) \leqslant 1$ for all j and k, we know that $\sum_{j=0}^{k-1} \delta(P^{(j,k)}) \leqslant N + m$. If $k > N + m$, define $M = [(k - N - 1)/m]$, where $[\cdot]$ represents the greatest integer function. Then

$$N \leqslant k - Mm - 1 < N + m,$$

and we can write

$$\sum_{j=0}^{k-1} \delta(P^{(j,k)}) = \sum_{j=0}^{k-Mm-1} \delta(P^{(j,k)}) + \sum_{j=k-Mm}^{k-(M-1)m-1} \delta(P^{(j,k)}) + \cdots$$

$$+ \sum_{j=k-m}^{k-1} \delta(P^{(j,k)}). \qquad (4.9)$$

In the first term of the right-hand side of (4.9) there are no more than $N + m$ summands, so

$$\sum_{j=0}^{k-Mm-1} \delta(P^{(j,k)}) \leqslant N + m.$$

In the second term of (4.9), note that $j \geqslant k - Mm > N$ so that inequality (4.8) can be used to get

$$\delta(P^{(j,k)}) = \delta(P^{(j,k-(M-1)m)} \cdot P^{(k-(M-1)m,k-(M-2)m)} \cdot \cdots \cdot P^{(k-m,k)})$$

$$\leqslant \delta(P^{(j,k-(M-1)m)}) \cdot \delta(P^{(k-(M-1)m,k-(M-2)m)}) \cdot \cdots \cdot \delta(P^{(k-m,k)})$$

$$\leqslant 1 \cdot \frac{3}{4} \cdot \cdots \cdot \frac{3}{4} = \left(\frac{3}{4}\right)^{M-1}.$$

Since this holds for $j = k - Mm, k - Mm + 1, \ldots, k - (M-1)m - 1$ we get

$$\sum_{j=k-Mm}^{k-(M-1)m-1} \delta\left(P^{(j,k)}\right) \leqslant m\left(\tfrac{3}{4}\right)^{M-1}.$$

If we apply the same reasoning to the remaining terms of (4.9) we get

$$\sum_{j=0}^{k-1} \delta\left(P^{(j,k)}\right) \leqslant (N+m) + m\left(\tfrac{3}{4}\right)^{M-1} + m\left(\tfrac{3}{4}\right)^{M-2} + \ldots + m\left(\tfrac{3}{4}\right)^{1} + m$$

$$= (N+m) + m\sum_{i=0}^{M-1}\left(\tfrac{3}{4}\right)^{i} < (N+m) + m\sum_{i=0}^{\infty}\left(\tfrac{3}{4}\right)^{i}$$

$$= N + m + m(4) = N + 5m.$$

Therefore if we take $D = N + 5m$, inequality (4.7) will be satisfied for all k. Since the hypotheses of Theorem V.4.4 are satisfied, the chain must be strongly ergodic. ▲

The method given above to prove Theorem V.4.5 was motivated by a desire to use Theorem V.4.4. Hence, the role of the left eigenvectors and the δ-coefficients became crucial and led in turn to the need for the lemmas preceeding the theorem. There does exist a more direct proof of Theorem V.4.5 that does not need Theorem V.4.4 or some of the lemmas. An outline of this proof will be given in the exercises. However, we believe that the longer proof given above is useful in that some of the lemmas are of interest by themselves and, furthermore, the proof provides a technique for finding the number D that is used in Theorem V.4.4. This technique is illustrated in the following example.

Example V.4.7. Let $\{X_n\}$ be a nonstationary Markov chain with transition matrices given by

$$P_1 = \begin{pmatrix} 1 & 0 \\ 0 & 1 \end{pmatrix}, \qquad P_n = \begin{bmatrix} \dfrac{1}{2} - \dfrac{1}{n} & \dfrac{1}{2} + \dfrac{1}{n} \\ 1 & 0 \end{bmatrix}$$

for $n = 2, 3, 4, \ldots$. It is easy to see that

$$P_n \to P = \begin{pmatrix} \tfrac{1}{2} & \tfrac{1}{2} \\ 1 & 0 \end{pmatrix}$$

as $n \to \infty$ and that this limiting matrix is weakly ergodic, so it follows from Theorem V.4.5 that the chain is strongly ergodic. On the other hand, if we wish to show strong ergodicity by using Theorem V.4.4, we first note that

$$\psi_n = \left(\frac{2n}{3n+2}, \frac{n+2}{3n+2} \right)$$

is a left eigenvector of P_n for $n = 1, 2, \ldots$. Also, $\psi_n \to \psi = (\frac{2}{3}, \frac{1}{3})$, where ψ is a left eigenvector for the limiting matrix P, so the first condition of Theorem V.4.4 is satisfied. Next we see that $\delta(P) = \frac{1}{2}$ and $\delta(P_n) = \frac{1}{2} + 1/n$ for $n \geq 2$ so, in the notation of the proof of Theorem V.4.5, $m = 1$ and $N = 4$. Therefore $D = N + 5m = 9$ will be such that $\sum_{j=0}^{k-1} \delta(P^{(j,k)}) \leq D$ for all k. This is easily seen since $M = k - N - 1$ and it follows that (4.9) becomes

$$\sum_{j=0}^{k-1} \delta(P^{(j,k)}) = \sum_{j=0}^{N} \delta(P^{(j,k)}) + \sum_{j=N+1}^{N+1} \delta(P^{(j,k)})$$

$$+ \sum_{j=N+2}^{N+2} \delta(P^{(j,k)}) + \ldots + \sum_{j=k-1}^{k-1} \delta(P^{(j,k)})$$

$$= \sum_{j=0}^{4} \delta(P^{(j,k)}) + \delta(P^{(N+1,k)}) + \delta(P^{(N+2,k)}) + \ldots + \delta(P^{(k-1,k)})$$

$$\leq 5 + \left(\tfrac{3}{4}\right)^{k-(N+1)} + \left(\tfrac{3}{4}\right)^{k-(N+2)} + \ldots + \left(\tfrac{3}{4}\right)^{k-(k-1)}$$

$$= 5 + \left(\tfrac{3}{4}\right)^{M} + \left(\tfrac{3}{4}\right)^{M-1} + \ldots + \left(\tfrac{3}{4}\right)^{1}$$

$$= 5 + \sum_{i=1}^{M} \left(\tfrac{3}{4}\right)^{i} < 5 + \sum_{i=1}^{\infty} \left(\tfrac{3}{4}\right)^{i} = 5 + 3 < D = 9.$$

Theorem V.4.5 may be regarded as a generalization of Corollary V.4.1 (or statistically speaking, we might say that Corollary V.4.1 is robust) in the following sense. If a stationary chain has a weakly ergodic transition matrix P, then the chain is strongly ergodic. If the assumption of stationarity is changed to a weaker assumption of nonstationarity and convergence of P_n to P in norm, the conclusion of Corollary V.4.1 still holds. Furthermore, the long run distribution depends only on P, the weakly ergodic limiting matrix. It is worth noting that while the long run distribution is independent of the rate at which $\{P_n\}$ converges to P, the rate of convergence might be of interest in certain situations. For example, the value of n such that $\| P_1 P_2 \ldots P_n - Q \| < \epsilon$ is certainly dependent on the rate of convergence. (A brief discussion of how the rate of convergence of P_n to

P affects the rate of convergence of $P_1P_2...P_n$ to Q is contained in Exercises 22 and 23.)

The hypotheses of Theorem V.4.5 required that the limiting matrix P be weakly ergodic. We might wonder whether the theorem would be true if P were not weakly ergodic. A little reflection will convince us that this is not true. For example, if P is not weakly ergodic, then $\{P_n\}$ defined by $P_n = P$ for all n will not be strongly ergodic, even though P_n converges to P (trivially). However, we can give examples where $\{P_n\}$ is strongly ergodic and converges to a matrix that is not weakly ergodic. Before giving such an example, we will consider once again Theorem V.4.3. Recall that the assumptions of this theorem are:

(i) $\{P_n\}$ is weakly ergodic.

(ii) $\sum_{n=1}^{\infty} \|\psi_n - \psi_{n+1}\| < \infty$. [This is (4.1) in the statement of Theorem V.4.3.]

We first note that for 2×2 matrices, weak ergodicity of $\{P_n\}$ depends on the rate of convergence to P when the limit, P, is *not* weakly ergodic.

If $\{P_n\}$ and $\{Q_n\}$ are defined by

$$P_n = \begin{bmatrix} 1 - \dfrac{1}{\sqrt{n}} & \dfrac{1}{\sqrt{n}} \\ 0 & 1 \end{bmatrix}, \qquad Q_n = \begin{bmatrix} 1 - \dfrac{1}{n^2} & \dfrac{1}{n^2} \\ 0 & 1 \end{bmatrix}, \qquad (4.10)$$

then both of these converge to the identity matrix, which is not weakly ergodic. Further, for every $n > 1$, $\|P_n - I\| > \|Q_n - I\|$, that is $\{P_n\}$ converges more slowly than $\{Q_n\}$. Also, since $\delta(P_n) = 1 - 1/\sqrt{n}$, we know that

$$\delta(P^{(k,n)}) \leqslant \prod_{j=k+1}^{n} \delta(P_j) = \prod_{j=k+1}^{n} \left(1 - \frac{1}{\sqrt{j}}\right) \to 0 \quad \text{as } n \to \infty,$$

so $\{P_n\}$ is weakly ergodic. On the other hand, for 2×2 stochastic matrices it can be shown that $\delta(PQ) = \delta(P)\delta(Q)$ (Exercise 10). Hence it follows that $\{Q_n\}$ is not weakly ergodic.

In order to show that $\{P_n\}$ as defined in (4.10) is strongly ergodic, it suffices to show that condition (4.1) of Theorem V.4.3 holds. It is easy to see that $\psi_n = (0, 1)$ for all n, so that condition (4.1) does hold. Hence $\{P_n\}$ is strongly ergodic even though P_n converges to a matrix that is not weakly ergodic.

The next example shows that although the rate of convergence of P_n to P seems to be a factor in the weak ergodicity of $\{P_n\}$, the rate may not be a factor in satisfying condition (4.1).

Example V.4.8. Let $\{P_n\}$ and $\{Q_n\}$ be defined by

$$P_n = \begin{bmatrix} \dfrac{1}{n} & 1-\dfrac{1}{n} \\ 1-\dfrac{2}{n} & \dfrac{2}{n} \end{bmatrix} \quad \text{for } n = 2,3,\ldots$$

$$Q_n = \begin{bmatrix} \dfrac{2}{n} & 1-\dfrac{2}{n} \\ 1-\dfrac{1}{n} & \dfrac{1}{n} \end{bmatrix} \quad \text{for } n \text{ even, } Q_n = P_n \text{ for } n \text{ odd.}$$

(For completeness we could take P_1 to be the identity). Clearly, both of these sequences converge to $P = \begin{pmatrix} 0 & 1 \\ 1 & 0 \end{pmatrix}$ and the rate of convergence is the same since $\|P_n - P\| = \|Q_n - P\|$. Further, the reader can verify that $\delta(P_n) = \delta(Q_n) = 1 - 3/n$, $n \geqslant 3$, and that both $\{P_n\}$ and $\{Q_n\}$ are weakly ergodic.

Now consider the behavior of the corresponding left eigenvectors. Let ψ_n and ϕ_n be the left eigenvectors corresponding to P_n and Q_n, respectively. Then

$$\psi_n = \left(\frac{n-2}{2n-3}, \frac{n-1}{2n-3} \right), \qquad n \geqslant 2$$

$$\phi_n = \left(\frac{n-1}{2n-3}, \frac{n-2}{2n-3} \right), \qquad n \text{ even, } \phi_n = \psi_n, n \text{ odd}.$$

Hence

$$\|\psi_{n+1} - \psi_n\| = \left| \frac{(n+1)-2}{2(n+1)-3} - \frac{n-2}{2n-3} \right| + \left| \frac{(n+1)-1}{2(n+1)-3} - \frac{n-1}{2n-3} \right|$$

$$= \frac{|(2n-3)(n-1) - (n-2)(2n-1)| + |(2n-3)(n) - (n-1)(2n-1)|}{(2n-1)(2n-3)}$$

$$= \frac{|2n^2 - 5n + 3 - 2n^2 + 5n - 2| + |2n^2 - 3n - 2n^2 + 3n - 1|}{(2n-1)(2n-3)}$$

$$= \frac{2}{(2n-1)(2n-3)} .$$

Clearly $\sum_{n=1}^{\infty} \| \psi_n - \psi_{n+1} \| < \infty$, that is condition (ii) holds, and we conclude that $\{P_n\}$ is strongly ergodic. On the other hand, for n odd,

$$\| \phi_n - \phi_{n+1} \| = \left| \frac{(n+1)-1}{2(n+1)-3} - \frac{(n-2)}{2n-3} \right| + \left| \frac{(n+1)-2}{2(n+1)-3} - \frac{(n-1)}{2n-3} \right|$$

$$= \frac{|(2n-3)n - (n-2)(2n-1)| + |(2n-3)(n-1) - (n-1)(2n-1)|}{(2n-1)(2n-3)}$$

$$= \frac{4(n-1)}{(2n-1)(2n-3)},$$

and the same expression will result for n even. In this case, $\sum_{n=1}^{\infty} \| \phi_n - \phi_{n+1} \| = \infty$, and so Theorem V.4.3 cannot be used to show that $\{Q_n\}$ is strongly ergodic. In summary, this example shows that the rate of convergence of P_n to P is not directly related to the convergence or divergence of $\sum_{n=1}^{\infty} \| \psi_n - \psi_{n+1} \|$.

The following theorem gives sufficient conditions for strong ergodicity of a special class of transition matrices.

Theorem V.4.6.

Let

$$P_n = \begin{pmatrix} \alpha_n & 1-\alpha_n \\ 1-\beta_n & \beta_n \end{pmatrix}$$

be a sequence of 2×2 stochastic matrices converging to $P = \begin{pmatrix} 0 & 1 \\ 1 & 0 \end{pmatrix}$. The sequence $\{P_n\}$ is strongly ergodic if

$$\sum_{n=1}^{\infty} (\alpha_n + \beta_n) = \infty \quad \text{and} \quad \sum_{n=1}^{\infty} |\alpha_n - \alpha_{n+1}| + |\beta_n - \beta_{n+1}| < \infty.$$

Proof. The essential ideas of the proof can be found in Example V.4.8. The details of the proof are left as an exercise. ▲

In view of Theorem V.4.6 and Example V.4.8 the following summary for 2×2 matrices converging to $\begin{pmatrix} 0 & 1 \\ 1 & 0 \end{pmatrix}$ can be given. If

$$P_n = \begin{pmatrix} \alpha_n & 1-\alpha_n \\ 1-\beta_n & \beta_n \end{pmatrix},$$

the rates at which α_n and β_n converge to zero are important in determining strong ergodicity but the rate alone is not sufficient. The condition $\sum_{n=1}^{\infty}|\alpha_n - \alpha_{n+1}| + |\beta_n - \beta_{n+1}| < \infty$ involves more than the rate at which α_n and β_n converge to zero, as Example V.4.8 shows. However, Theorem V.4.6 does yield conditions for strong ergodicity that are reasonably easy to check in the case of 2×2 matrices converging to $\begin{pmatrix} 0 & 1 \\ 1 & 0 \end{pmatrix}$. When strong ergodicity does hold, the limit matrix, Q, is easy to find (Exercise 8). We next raise the natural question of what happens in higher dimensions. If $\{P_n\}$ is a sequence of $k \times k$ stochastic matrices converging to the periodic stochastic matrix, what criteria on the entries converging to zero would guarantee strong ergodicity? As far as the authors know this is an open question.

The final case to consider for 2×2 stochastic matrices is the case where $P_n \to \begin{pmatrix} 1 & 0 \\ 0 & 1 \end{pmatrix}$. This is in some sense the most interesting case since the limit stochastic matrix, I, does not have a unique invariant probability vector (i.e., all probability vectors are left eigenvectors of the identity matrix). Hence there is no natural constant matrix, Q, to which the nonstationary chain should converge. For this case the left eigenvector of

$$P_n = \begin{pmatrix} 1 - \alpha_n & \alpha_n \\ \beta_n & 1 - \beta_n \end{pmatrix}$$

is

$$\left(\frac{\beta_n}{\alpha_n + \beta_n}, \frac{\alpha_n}{\alpha_n + \beta_n} \right)$$

and so condition (4.1) of Theorem V.4.3 becomes

$$\sum_{n=1}^{\infty} \left| \frac{\beta_n}{\alpha_n + \beta_n} - \frac{\beta_{n+1}}{\alpha_{n+1} + \beta_{n+1}} \right| + \left| \frac{\alpha_n}{\alpha_n + \beta_n} - \frac{\alpha_{n+1}}{\alpha_{n+1} + \beta_{n+1}} \right| < \infty. \quad (4.11)$$

Now if $\sum_{n=1}^{\infty}(\alpha_n + \beta_n) = \infty$ and (4.11) holds, the limit of the strongly ergodic chain has identical rows equal to

$$\lim_{n \to \infty} \left(\frac{\beta_n}{\alpha_n + \beta_n}, \frac{\alpha_n}{\alpha_n + \beta_n} \right)$$

by Theorem V.4.3. We see that in this case $Q = \lim_{n \to \infty} P_1 P_2 \cdots P_n$ is completely determined by α_n and β_n while in the previous cases Q was determined by the matrix $P = \lim_{n \to \infty} P_n$.

As in the periodic case, there is the open question of higher-dimensional stochastic matrices converging to the identity matrix. The final question would be to consider $P_n \to P$ in norm where P is both reducible and periodic.

Example V.4.9. Let $\{X_n\}$ be a nonstationary Markov chain with transition matrices

$$
P_n = \begin{pmatrix} 1 - \dfrac{1}{n} & \dfrac{1}{n} \\[2mm] \dfrac{2}{n} & 1 - \dfrac{2}{n} \end{pmatrix} \quad \text{for } n \geqslant 2.
$$

and $P_1 = \begin{pmatrix} 1 & 0 \\ 0 & 1 \end{pmatrix}$. Now it is easy to see that $\lim_{n \to \infty} P_n = I$, so we are in the case discussed above. For this example $\alpha_n = 1/n$ and $\beta_n = 2/n$ so

(i) $\displaystyle\sum_{n=1}^{\infty} (\alpha_n + \beta_n) = \infty$ and

(ii) $\displaystyle\sum_{n=1}^{\infty} \left| \frac{\beta_n}{\alpha_n + \beta_n} - \frac{\beta_{n+1}}{\alpha_{n+1} + \beta_{n+1}} \right| + \left| \frac{\alpha_n}{\alpha_n + \beta_n} - \frac{\alpha_{n+1}}{\alpha_{n+1} + \beta_{n+1}} \right| = 0.$

Hence the chain is strongly ergodic with limiting matrix

$$
Q = \begin{pmatrix} \frac{2}{3} & \frac{1}{3} \\[1mm] \frac{2}{3} & \frac{1}{3} \end{pmatrix}.
$$

We conclude Section V.4 by considering whether strong ergodicity implies convergence of the corresponding left eigenvectors. That is, we consider whether the convergence of the left eigenvectors is necessary for strong ergodicity. Unfortunately, this is not true in general as the following example shows.

Example V.4.10. Let $\{X_n\}$ be a nonstationary Markov chain with transition matrices

$$
P_{2n} = \begin{pmatrix} 1 - \dfrac{1}{2n} & \dfrac{1}{2n} \\[2mm] 0 & 1 \end{pmatrix} \quad \text{and} \quad P_{2n-1} = \begin{pmatrix} \frac{1}{2} & \frac{1}{2} \\[1mm] \frac{1}{2} & \frac{1}{2} \end{pmatrix} \quad \text{for } n = 1, 2, 3, \ldots .
$$

Since $\psi_{2n} = (0, 1)$ and $\psi_{2n-1} = (\frac{1}{2}, \frac{1}{2})$ the corresponding left eigenvectors do not converge. However,

$$P_1 P_2 = \begin{pmatrix} \frac{1}{4} & \frac{3}{4} \\ \frac{1}{4} & \frac{3}{4} \end{pmatrix}, \qquad P_1 P_2 P_3 = \begin{pmatrix} \frac{1}{2} & \frac{1}{2} \\ \frac{1}{2} & \frac{1}{2} \end{pmatrix}$$

and in general

$$P_1 P_2 \cdots P_{2n} = \begin{bmatrix} \dfrac{1}{2} - \dfrac{1}{4n} & \dfrac{1}{2} + \dfrac{1}{4n} \\ \dfrac{1}{2} - \dfrac{1}{4n} & \dfrac{1}{2} + \dfrac{1}{4n} \end{bmatrix} \quad \text{and} \quad P_1 P_2 \cdots P_{2n+1} = \begin{pmatrix} \frac{1}{2} & \frac{1}{2} \\ \frac{1}{2} & \frac{1}{2} \end{pmatrix}$$

so the chain is strongly ergodic.

The following theorem gives conditions in addition to strong ergodicity which guarantee the convergence of the left eigenvectors.

Theorem V.4.7.

Let $\{P_n\}$ be a strongly ergodic sequence of stochastic matrices. If there exists an integer k and a real number β such that

$$\delta\left(P_n^k\right) \leqslant \beta < 1$$

for all n, then $\{\psi_n\}$ converges in norm to ψ where ψ is the common row of the constant matrix $Q = \lim_{n \to \infty} P^{(1,n)}$.

Proof. First note that $\delta(P_n^k) \leqslant \beta < 1$ for fixed n implies that P_n considered as a stationary chain is weakly ergodic and hence there is a *unique* left eigenvector ψ_n with $\|\psi_n\| = 1$ (see Theorem VI.4.2). Define Q_n to be a constant matrix with each row equal to ψ_n.
In the following we take $m = 0$, but the same proof holds for arbitrary m. Define

$$E_n = P^{(0,n)} - P^{(0,n-1)}$$

and

$$D_n = Q_n - P^{(0,n-1)}. \tag{4.12}$$

Since

$$\|\psi_n - \psi\| = \|Q_n - Q\| \leqslant \|Q_n - P^{(0,n-1)}\| + \|P^{(0,n-1)} - Q\|$$

and since $\|P^{(0,n-1)}-Q\|\to 0$ as $n\to\infty$ by assumption, it suffices to show that $\|Q_n - P^{(0,n-1)}\| = \|D_n\| \to 0$ as $n\to\infty$. Multiplying both sides of (4.12) by P_n and using the fact that $Q_n P_n = Q_n$, we get

$$D_n P_n = Q_n - P^{(0,n)} = D_n - E_n.$$

Hence $D_n = D_n P_n + E_n$, so substituting for D_n on the right-hand side we get $D_n = (D_n P_n + E_n)P_n + E_n = D_n P_n^2 + E_n P_n + E_n$.
Continuing this we get

$$D_n = D_n P_n^M + \sum_{j=0}^{M-1} E_n P_n^j$$

where P_n^0 is the identity matrix. Hence

$$\|D_n\| \le \|D_n P_n^M\| + \left\| \sum_{j=0}^{M-1} E_n P_n^j \right\|.$$

Now D_n and E_n both have rows that sum to zero, so using Lemma V.2.4 we get

$$\|D_n\| \le 2\delta(P_n^M) + \|E_n\| \sum_{j=0}^{M-1} \delta(P_n^j).$$

Let $\epsilon > 0$ be given. Choose $M = M(\epsilon)$ so that $\delta(P_n^M) < \epsilon/4$. This is possible since $\delta(P_n^M) \le \{\delta(P_n^k)\}^{[M/k]} \le \beta^{[M/k]}$ (where $[\cdot]$ represents the greatest integer function). Now with M fixed, choose $N = N(\epsilon, M)$ such that $\|E_n\| \le \epsilon/2M$ for all $n \ge N$. This is possible since strong ergodicity implies that $\lim_{n\to\infty}\|E_n\| = 0$. With this choice of N we have $\|D_n\| \le \epsilon$ for $n \ge N$. ▲

In reviewing the examples given in this section, we see that if $\{P_n\}$ is a sequence of transition matrices defining a weakly ergodic nonstationary Markov chain, then while there is some relationship between convergence of the left eigenvectors of $\{P_n\}$ and strong ergodicity of the corresponding chain, this relationship is not direct. There is still a question as to whether or not some condition on the eigenvectors of $\{P_n\}$ is necessary and sufficient for a weakly ergodic chain to be strongly ergodic. At this time, we do not know the answer to this question.

EXERCISES

1. Consider a nonstationary Markov chain with transition matrices

$$P_{3n} = \begin{bmatrix} 1 - \dfrac{1}{3n} & 0 & \dfrac{1}{3n} \\[2mm] \dfrac{1}{3n} & 1 - \dfrac{1}{3n} & 0 \\[2mm] 0 & \dfrac{1}{3n} & 1 - \dfrac{1}{3n} \end{bmatrix},$$

$$P_{3n+1} = \begin{bmatrix} 0 & 1 & 0 \\ 0 & 0 & 1 \\ 1 & 0 & 0 \end{bmatrix}, \qquad P_{3n+2} = \begin{bmatrix} 1 & 0 & 0 \\ 0 & 1 & 0 \\ 0 & 0 & 1 \end{bmatrix}.$$

Determine whether or not this chain is strongly ergodic.

2. Show that the chain defined in Example V.4.1 is weakly ergodic and $\lim_{n \to \infty} \psi_n = (\tfrac{1}{2}, \tfrac{1}{2})$.

3. Let

$$P_n = \begin{bmatrix} \dfrac{1}{n} & 1 - \dfrac{1}{n} \\[2mm] 1 & 0 \end{bmatrix}$$

for $n = 1, 2, 3, \ldots$. Show that the corresponding nonstationary chain is strongly ergodic using Theorem V.4.3. Show that the hypotheses of Theorem V.4.4 are not satisfied for this chain.

4. Show that the strong ergodicity of the chain defined in Example V.4.5 cannot be established using Theorem V.4.3.

5. Show that the chain given in Example V.4.4 is strongly ergodic using Theorem V.4.4.

6. Let $\{X_n\}$ be a nonstationary Markov chain with transition matrices

$$P_n = \begin{bmatrix} 1 - \dfrac{1}{n} & \dfrac{1}{n} \\[2mm] \dfrac{4}{n} - \delta_n & 1 - \dfrac{4}{n} + \delta_n \end{bmatrix} \qquad \text{for } n \geq 4$$

and $P_n = I$ for $n < 4$. Find conditions on δ_n that guarantee $\{P_n\}$ is strongly ergodic. Under these conditions find $\lim_{n \to \infty} P_1 P_2 \cdots P_n$.

†7. Find a sequence, $\{P_n\}$, of 2×2 stochastic matrices such that

$\lim_{n \to \infty} P_n = I$, $\{P_n\}$ is strongly ergodic, and $\lim_{n \to \infty} P_1 P_2 \cdots P_n = (1/e, 1 - 1/e)$.

8.   Show that if $\{P_n\}$ satisfies the hypotheses of Theorem V.4.3 and

$$P_n \to \begin{pmatrix} 0 & 1 \\ 1 & 0 \end{pmatrix}, \text{ then } P_1 P_2 \cdots P_n \to \begin{pmatrix} \frac{1}{2} & \frac{1}{2} \\ \frac{1}{2} & \frac{1}{2} \end{pmatrix}.$$

9.   Discuss Lemma V.4.4 for the two cases

$$\text{(a)} \quad \lim_{n \to \infty} P_n = \begin{pmatrix} 0 & 1 \\ 1 & 0 \end{pmatrix},$$

$$\text{(b)} \quad \lim_{n \to \infty} P_n = \begin{pmatrix} 1 & 0 \\ 0 & 1 \end{pmatrix}.$$

*10.   (a)   Show that for 2×2 stochastic matrices, $\delta(PQ) = \delta(P) \cdot \delta(Q)$.
       (b)   Use this fact to show that the sequence of stochastic matrices, $\{Q_n\}$, given in (4.10) is not weakly ergodic.

*11.   Show that if A is a constant matrix and B is stochastic, then $BA = A$.

12.   Prove Corollary V.4.1 using Theorem V.4.4.

*13.   Prove Theorem V.4.1.

14.   Show that if $\sum_{j=1}^{k} \delta(P^{(j,k)}) \leqslant D$ for all k, then $\delta(P^{(m,k)}) \to 0$ as $k \to \infty$ for all m. Show that the converse is false.

15.   Complete the induction proof of Lemma V.4.3.

16.   Prove Lemma V.4.5.

17.   Prove Theorem V.4.6.

18.   Show that every finite stochastic matrix, P, has at least one invariant left probability vector, ψ.

19.   Show that if $\sum_{j=1}^{\infty} \|\psi_{j+1} - \psi_j\| < \infty$, then $\{\psi_j\}$ is Cauchy in norm and hence $\lim_{n \to \infty} \psi_n = \psi$ exists.

†20.   Use Lemmas V.2.5, V.4.5 and Exercise 11 to prove Theorem V.4.5. Hint:

$$\| P^{(1,n)} - Q \| \leqslant \| P^{(1, n-N)} P^{(n-N, n)} - P^{(1, n-N)} P^N \|$$

$$+ \| P^{(1, n-N)} P^N - P^{(1, n-N)} Q \|.$$

21.   Show that for a 2×2 stochastic matrix P the absolute value of the non-unit eigenvalue for P is equal to $\delta(P)$.

†22.   Let P be a stochastic matrix corresponding to a weakly ergodic Markov chain. Show that $\|P^n - Q\|$ goes to zero at a geometric rate. (That is show that there exists $\beta < 1$ such that $\|P^n - Q\| < K\beta^n$

for some constant K and for all n.) Hint: Recall that $QP^{n-1}=Q$ and use Lemma V.2.4. (For further results on the rate at which $x_0 P^n$ converges to π, see Pitman (1974).)

†23.   Assume $P_n \to P$ at such a rate that $n^\alpha \|P_n - P\| \to 0$ as $n \to \infty$ for some $\alpha > 0$. Show that if P is weakly ergodic, $n^\alpha \|P_1 P_2 \cdots P_n - Q\| \to 0$ as $n \to \infty$. Hint:

$$n^\alpha \|P_1 P_2 \cdots P_n - Q\| \le n^\alpha \big[\|P_1 P_2 \ldots P_n - P_1 P_2 \ldots P_{[n/2]} P^{n-[n/2]}\|$$

$$+ \|P_1 P_2 \ldots P_{[n/2]} P^{n-[n/2]} - Q\| \big]$$

and

$$\|P_{[n/2]+1} P_{[n/2]+2} \ldots P_n - P^{n-[n/2]}\| \le \sum_{k=0}^{[n/2]} \|P_{n-k} - P\| \delta(P^k).$$

†24.   From both the algebraic approach to Markov chains and from Exercise 22, we know that

$$P = \begin{bmatrix} \frac{1}{4} & \frac{3}{4} & 0 \\ 0 & 0 & 1 \\ 0 & \frac{1}{2} & \frac{1}{2} \end{bmatrix}$$

has the property that P^n converges to Q at a geometric rate. Which approach yields the faster rate in this case? (See Equation 2.4 of Chapter IV and Exercise 22 above.)

†25.   Let $\{X_n\}$ be a nonstationary Markov chain with transition matrices

$$P_n = \begin{bmatrix} \frac{1}{n} & 1-\frac{1}{n} & 0 & 0 & 0 & \cdot & \cdot & \cdot \\ \frac{1}{n} & 0 & 1-\frac{1}{n} & 0 & 0 & \cdot & \cdot & \cdot \\ \frac{1}{n} & 0 & 0 & 1-\frac{1}{n} & 0 & \cdot & \cdot & \cdot \\ \cdot & \cdot & \cdot & \cdot & \cdot & & & \\ \cdot & \cdot & \cdot & \cdot & \cdot & \cdot & \cdot & \cdot \\ \cdot & \cdot & \cdot & \cdot & \cdot & \cdot & \cdot & \cdot \end{bmatrix}$$

(a)   Show that $\{P_n\}$ is weakly ergodic.
(b)   Is $\{P_n\}$ uniformly weakly ergodic? (That is, does $\delta(P^{(m,m+n)}) \to 0$ as $n \to \infty$ uniformly in m?)
(c)   Is $\{P_n\}$ strongly ergodic?
(d)   Show that the mean return time to state 1 is infinite.

SECTION 5   **C-STRONG ERGODICITY**

In Section V.4 we considered various theorems that gave sufficient conditions for strong ergodicity. In particular, several results were given for Markov chains with the property that $\lim_{n\to\infty} P_n = P$. However only 2×2 matrices were considered in cases where P was reducible or periodic. In this section we prove some ergodic theorems in the case where $\lim_{n\to\infty} P_n = P$ and P is periodic. [See Bowerman, David, and Isaacson (1975).]

We see immediately from Chapter III that if P is periodic, the sequences $\{p_{ii}^{(0,n)}\}$ can oscillate as $n\to\infty$. For example if $P_n \equiv P = \begin{pmatrix} 0 & 1 \\ 1 & 0 \end{pmatrix}$ then $\lim_{n\to\infty} p_{11}^{(0,n)}$ does not exist. However, the stationary periodic chains of Chapter III do exhibit the following two types of ergodicity. If the transition matrix, P, has period d, then $\lim_{n\to\infty} P^{nd+m}$ exists for $m = 0, 1, \ldots, d-1$ and $\lim_{n\to\infty}(1/n)\sum_{k=1}^{n} P^k$ exists. In view of these results we will consider whether analogous results hold for nonstationary chains for which $\lim_{n\to\infty} P_n = P$ and P is irreducible and periodic. That is, we will look for convergence of subsequences and/or Cesaro averages of $P_1 P_2 \cdots P_n$.

Definition V.5.1.   *A sequence of stochastic matrices, $\{P_n\}$, is said to converge in the Cesaro sense to the matrix Q if*

$$\lim_{n\to\infty} \left\| \frac{1}{n} \sum_{k=1}^{n} P_1 P_2 \ldots P_k - Q \right\| = 0.$$

We also will refer to this as convergence of the Cesaro averages or C-strong ergodicity.

We begin by showing that if $\lim_{n\to\infty} P_1 P_2 \ldots P_{nd+m}$ exists for some fixed $d \geqslant 1$ and all m with $0 \leqslant m < d$, then $\{P_n\}$ converges in the Cesaro sense. That is if the subsequences converge, then the Cesaro averages converge. (For notational convenience we will sometimes take $1 \leqslant m \leqslant d$.)

Lemma V.5.1.   Let $\{P_n\}$ be a sequence of stochastic matrices such that $\lim_{n\to\infty} P_1 P_2 \ldots P_{nd+m} = Q_m$ for some fixed d and all m with $1 \leqslant m \leqslant d$. Then $\lim_{n\to\infty}(1/n)\sum_{k=1}^{n} P_1 P_2 \ldots P_k$ exists and equals $Q = (1/d)[Q_1 + Q_2 + \ldots + Q_d]$. (Note that the limit in both cases is a limit in the norm, $\|\cdot\|$.)

Proof.   We first write

$$\frac{1}{n} \sum_{k=1}^{n} P_1 P_2 \ldots P_k = \frac{1}{n} \sum_{k=1}^{d[n/d]} P_1 P_2 \ldots P_k + \frac{1}{n} \sum_{k=d[n/d]+1}^{n} P_1 P_2 \ldots P_k \quad (5.1)$$

where, as usual, $[\cdot]$ represents the greatest integer function and the second term is defined to be zero if $d[n/d]+1 > n$. It follows that

$$\left\| \frac{1}{n} \sum_{k=d[n/d]+1}^{n} P_1 P_2 \dots P_k \right\| \leqslant \frac{d}{n},$$

which goes to zero as $n \to \infty$. Hence it is sufficient to show that

$$\lim_{n \to \infty} \left\| \frac{1}{n} \sum_{k=1}^{d[n/d]} P_1 P_2 \dots P_k - Q \right\| = 0.$$

Note that

$$\sum_{k=1}^{d[n/d]} P_1 P_2 \dots P_k = \sum_{m=1}^{d} \sum_{k=0}^{[n/d]-1} P_1 P_2 \dots P_{kd+m},$$

so we get

$$\lim_{n \to \infty} \frac{1}{n} \sum_{k=1}^{d[n/d]} P_1 P_2 \dots P_k = \lim_{n \to \infty} \frac{1}{n} \sum_{m=1}^{d} \sum_{k=0}^{[n/d]-1} P_1 P_2 \dots P_{kd+m}$$

$$= \sum_{m=1}^{d} \lim_{n \to \infty} \frac{[n/d]}{n} \cdot \frac{1}{[n/d]} \sum_{k=0}^{[n/d]-1} P_1 P_2 \dots P_{kd+m}. \quad (5.2)$$

From Chapter I we know that if a sequence converges, then the Cesaro averages converge to the same limit. We also have that $\lim_{n \to \infty}([n/d]/n) = 1/d$ so

$$\lim_{n \to \infty} \frac{1}{n} \sum_{k=1}^{d[n/d]} P_1 P_2 \dots P_k = \sum_{m=1}^{d} \frac{1}{d} \cdot Q_m = Q. \quad \blacktriangle$$

Recall that we are interested in nonstationary Markov chains for which the transition matrices, $\{P_n\}$, converge in norm to P where P is periodic of period d. From the results obtained in the case of stationary periodic chains we conjecture that $\lim_{k \to \infty} P_1 P_2 \dots P_{kd+m}$ will exist in norm and hence from Lemma V.5.1 the Cesaro averages will also converge. Unfortunately the following example shows that subsequences of the form $P_1 P_2 \dots P_{nd+m}$ may not converge even though $\lim_{n \to \infty} P_n = P$ is periodic of period d.

Example V.5.1. In this example we construct a nonstationary chain that has the property that $\lim_{n\to\infty} P_n = P$ and P is periodic of period d, but for which subsequences of the form $P_1 P_2 \dots P_{nd+m}$ fail to converge as $n \to \infty$.

Define $P_n = \begin{pmatrix} 0 & 1 \\ 1 & 0 \end{pmatrix}$ for n odd. For n even define P_n to be either S_n or T_n where

$$S_n = \begin{bmatrix} \dfrac{1}{n} & 1 - \dfrac{1}{n} \\ 1 & 0 \end{bmatrix}, \qquad T_n = \begin{bmatrix} 0 & 1 \\ 1 - \dfrac{1}{n} & \dfrac{1}{n} \end{bmatrix}.$$

Of course we will not arbitrarily choose P_n to be either S_n or T_n, but we will have to make the choice rather judiciously in order to construct a chain with the desired properties. However, no matter what choice is made it will be true that $\lim_{n\to\infty} P_n = \begin{pmatrix} 0 & 1 \\ 1 & 0 \end{pmatrix}$. Furthermore we can see that the chain will be weakly ergodic since $\delta(P_{2n}) = 1 - 1/2n$, independently of the choice made at even times. (But an attempt to show strong ergodicity using Theorem V.4.3 will fail.)

We will construct the chain so that $\mathbf{f}^{(2n)} = (1,0)P_1 P_2 \dots P_{2n}$ fails to converge in norm as $n \to \infty$. [Likewise $\mathbf{f}^{(2n+1)} = (1,0)P_1 P_2 \dots P_{2n+1}$ fails to converge in norm.] Consider $\mathbf{f}^{2n} = (0,1)(P_1 P_2)(P_3 P_4) \dots (P_{2n-1} P_{2n})$. If the matrix at every even time were taken to be S_n, the chain with transition matrices, $R_n = P_{2n-1} P_{2n}$ would be strongly ergodic with $\lim_{n\to\infty} R_n = \begin{pmatrix} 1 & 0 \\ 1 & 0 \end{pmatrix}$ (Use Theorem V.4.3). Similarly, if the matrix at time $2n$ were taken to be T_n, $n = 1, 2, 3, \dots$, the chain with transition matrices, R_n would be strongly ergodic with $\lim_{n\to\infty} R_n = \begin{pmatrix} 0 & 1 \\ 0 & 1 \end{pmatrix}$. By appropriate choice of P_n at each even time, an oscillating sequence, $\{\mathbf{f}^{(2n)}\}$, can be constructed. In particular the first coordinate of $\mathbf{f}^{(2n)}$ can be made less than $\frac{1}{4}$ and greater than $\frac{3}{4}$ infinitely often. This can be done as follows:

(i) $\mathbf{f}^{(0)} = (1,0)$ so we choose $P_k = T_k$ for $k = 2, 4, \dots, 2N_1$ so that $\mathbf{f}^{(2N_1)} = (1,0)P_1 P_2 \dots P_{2N_1 - 1} P_{2N_1}$ has a first coordinate less than $\frac{1}{4}$. This will happen in a finite number of steps since $\mathbf{f}^{(2n)} \to (0,1)$ when $P_k = T_k$ for all even k.

(ii) Once the first coordinate is less than $\frac{1}{4}$, switch to $P_k = S_k$ for $k = 2N_1 + 2, 2N_1 + 4, \dots, 2N_2$ where N_2 is such that the first coordinate of $\mathbf{f}^{(2N_2)}$ is greater than $\frac{3}{4}$. This again will occur in a finite number of steps since $(P_{2N_1+1} S_{2N_1+2})(P_{2N_1+3} S_{2N_1+4}) \dots (P_{2m-1} S_{2m})$ is strongly ergodic. That is

$$\left[(1,0)(P_1 T_2)(P_3 T_4) \dots (P_{2N_1 - 1} T_{2N_1}) \right] (P_{2N_1 + 1} S_{2N_1 + 2})$$

$$\times (P_{2N_1 + 3} S_{2N_1 + 4}) \dots (P_{2m-1} S_m) \to (1,0)$$

where $(1,0)(P_1 T_2)(P_3 T_4)\ldots(P_{2N_1-1} T_{2N_1})$ plays the role of the starting vector.

(iii) Switching back to choosing T_n at the even times we again get the first coordinate of $\mathbf{f}^{(2N_3)}$ less than $\frac{1}{4}$ where $N_3 > N_2$.

(iv) Continuing this procedure for choosing P_{2n} we get oscillation in the first coordinate of $\mathbf{f}^{(2n)}$ so that the subsequence $(1,0)(P_1 P_2)\cdots$ $(P_{2n-1} P_{2n})$ does *not* converge as $n \to \infty$.

All is not lost, however, since it is still possible that the Cesaro averages will converge. Theorem V.5.1 does in fact give sufficient conditions for convergence of the Cesaro averages, but before proving this theorem we need some definitions and lemmas.

Definition V.5.2. *Let P be an irreducible stochastic matrix of period d. Define $P^{d(i)}$ to be the irreducible aperiodic stochastic matrix in P^d which applies to the ith cycle of P. (The fact that $P^{d(i)}$ is irreducible and aperiodic is left as an exercise.)*

Example V.5.2. Consider the stochastic matrix

$$P = \begin{bmatrix} 0 & \frac{1}{2} & 0 & \frac{1}{3} & \frac{1}{6} \\ \frac{1}{2} & 0 & \frac{1}{2} & 0 & 0 \\ 0 & \frac{1}{2} & 0 & 0 & \frac{1}{2} \\ \frac{1}{3} & 0 & \frac{2}{3} & 0 & 0 \\ 1 & 0 & 0 & 0 & 0 \end{bmatrix}.$$

It is easy to see that P has period two with cycles $D_0 = \{1,3\}$ and $D_1 = \{2,4,5\}$. Calculating P^2 we get

$$P^2 = \begin{bmatrix} \frac{19}{36} & 0 & \frac{17}{36} & 0 & 0 \\ 0 & \frac{1}{2} & 0 & \frac{1}{6} & \frac{1}{3} \\ \frac{3}{4} & 0 & \frac{1}{4} & 0 & 0 \\ 0 & \frac{1}{2} & 0 & \frac{1}{9} & \frac{7}{18} \\ 0 & \frac{1}{2} & 0 & \frac{1}{3} & \frac{1}{6} \end{bmatrix}.$$

Using the notation of Definition V.5.2 we get

$$P^{d(0)} = \begin{pmatrix} \frac{19}{36} & \frac{17}{36} \\ \frac{3}{4} & \frac{1}{4} \end{pmatrix} \quad \text{and} \quad P^{d(1)} = \begin{pmatrix} \frac{1}{2} & \frac{1}{6} & \frac{1}{3} \\ \frac{1}{2} & \frac{1}{9} & \frac{7}{18} \\ \frac{1}{2} & \frac{1}{3} & \frac{1}{6} \end{pmatrix}.$$

Note that the stochastic matrices $P^{d(0)}$ and $P^{d(1)}$ given in Example V.5.2 are weakly ergodic. In fact, for finite irreducible stochastic matrices of period d we automatically have for $i = 0, 1, \ldots, d-1$ that $P^{d(i)}$ is weakly ergodic since it is irreducible and aperiodic. However, for infinite chains the weak ergodicity of $P^{d(i)}$ does not follow from irreducibility and aperiodicity. (See Example V.3.4.) Hence in the following lemmas and theorems we will explicitly assume that $P^{d(i)}$ is weakly ergodic for $i = 0, 1, \ldots, d-1$.

Lemma V.5.2. Let $\{X_n\}$ be a nonstationary Markov chain with transition matrices $\{P_n\}$. Let $\lim_{n \to \infty} P_n = P$ where P is irreducible and periodic of period d. Assume $P^{d(i)}$ is weakly ergodic for $i = 0, 1, \ldots, d-1$. Then for each $m \geq 0$, we have

$$\lim_{n \to \infty} \left\| \frac{1}{d} \sum_{l=0}^{d-1} P^{(m, m+n+l)} - Q \right\| = 0.$$

The matrix Q is constant with each row equal to the left invariant probability vector for P. That is if ψ is a row of Q, then $\psi P = \psi$, $\sum_{i \in S} \psi_i = 1$, and $\psi_i > 0$ for all $i \in S$.

Proof. We begin this proof by considering the stationary Markov chain with transition matrix P. Since $P^{d(i)}$ is assumed to be weakly ergodic, we have $\lim_{n \to \infty} P^{nd} = Q_0$ exists in norm. In particular, each $P^{d(i)}$ converges in norm to a constant matrix and the remaining entries of P^{nd} are all zeros.

It follows that $\lim_{n \to \infty} P^{nd+l} = Q_l = Q_0 \cdot P^l$ for $l = 0, 1, \ldots, d-1$ (Exercise 7). Hence

$$\left\| \frac{1}{d} \sum_{l=0}^{d-1} P^{nd+l} - Q \right\| \to 0 \quad \text{as } n \to \infty \quad \text{where } Q = \frac{1}{d} \sum_{l=0}^{d-1} Q_l. \tag{5.3}$$

We next show that Q is invariant under P.

$$\left(\frac{1}{d} \sum_{l=0}^{d-1} Q_l \right) P = \left(\frac{1}{d} \sum_{l=0}^{d-1} Q_0 P^l \right) P = \frac{1}{d} \sum_{l=0}^{d-1} Q_0 P^{l+1}.$$

Using the fact that $Q_0 P^d = Q_0$ we get the desired result. It is easy to see that $\|Q\| = 1$ (Exercise 8). However the irreducible stochastic matrix P has a unique invariant probability vector ψ so each row of Q must be ψ.

Now consider the nonstationary Markov chain where $\lim_{n \to \infty} P_n = P$ and

P is irreducible and of period d. We have

$$\left\|\frac{1}{d}\sum_{l=0}^{d-1}P^{(m,m+n+l)}-Q\right\|=\left\|P^{(m,m+n-Md)}\left(\frac{1}{d}\sum_{l=0}^{d-1}P^{(m+n-Md,m+n+l)}-Q\right)\right\|$$

$$(5.4)$$

since Q is a constant matrix and $P^{(m,m+n-Md)}$ is stochastic. Add and subtract $P^{(m,m+n-Md)}[(1/d)\sum_{l=0}^{d-1}P^{Md+l}]$ to (5.4) and get

$$\left\|P^{(m,m+n-Md)}\left[\frac{1}{d}\sum_{l=0}^{d-1}\left(P^{(m+n-Md,m+n+l)}-P^{Md+l}\right)\right]\right.$$

$$\left.+P^{(m,m+n-Md)}\left(\frac{1}{d}\sum_{l=0}^{d-1}P^{Md+l}-Q\right)\right\|$$

$$\leqslant\frac{1}{d}\sum_{l=0}^{d-1}\left\|P^{(m+n-Md,m+n+l)}-P^{Md+l}\right\|$$

$$+\left\|\frac{1}{d}\sum_{l=0}^{d-1}P^{Md+l}-Q\right\|.$$

$$(5.5)$$

Let $\epsilon>0$ be given. Choose M sufficiently large so that by (5.3) we have

$$\left\|\frac{1}{d}\sum_{l=0}^{d-1}P^{Md+l}-Q\right\|<\frac{\epsilon}{2}.$$

For this fixed M and ϵ it is possible to choose $N=N(\epsilon,M)$ so that $\|P^{(m+n-Md,m+n+l)}-P^{Md+l}\|<\epsilon/2d$ for all $n>N$ and for $l=0,1,\ldots,d-1$. This can be done by Lemma V.4.5. Hence for $n>N$ we have

$$\left\|\frac{1}{d}\sum_{l=0}^{d-1}P^{(m,m+n+l)}-Q\right\|<\epsilon$$

and the proof is complete.   ▲

Lemma V.5.3. Let $\{P_n\}$ be a sequence of stochastic matrices such that for all m,

$$\left\|\frac{1}{d}\sum_{l=0}^{d-1}P^{(m,m+n+l)}-Q\right\|\to0 \quad \text{as } n\to\infty.$$

Then

$$\left\| \frac{1}{n} \sum_{t=1}^{n} P^{(m,m+t)} - Q \right\| \to 0 \quad \text{as } n \to \infty.$$

Proof. Write $(1/n)\sum_{t=1}^{n} P^{(m,m+t)}$ in the form

$$\frac{1}{n} \sum_{k=0}^{[n/d]-1} \sum_{l=0}^{d-1} P^{(m,m+kd+l+1)} + \frac{1}{n} \sum_{t=d[n/d]+1}^{n} P^{(m,m+t)} \qquad (5.6)$$

where the second term is zero if $d[n/d]+1$ exceeds n. Rewrite the first term of (5.6) as

$$\frac{d}{n} \sum_{k=0}^{[n/d]-1} \left(\frac{1}{d} \sum_{l=0}^{d-1} P^{(m,m+kd+l+1)} \right).$$

It follows from Exercise 12 of Section I.4 that this converges to Q as $n \to \infty$. Hence it is sufficient to show that

$$\lim_{n \to \infty} \frac{1}{n} \sum_{t=d[n/d]+1}^{n} P^{(m,m+t)} = 0,$$

but this is obvious since the sum contains at most d terms each of which has norm one. That is

$$\left\| \frac{1}{n} \sum_{t=d[n/d]+1}^{n} P^{(m,m+t)} \right\| \leqslant \frac{d}{n}.$$

This completes the proof of the lemma. ▲

We now give the main theorem on C-strong ergodicity. The proof is a straightforward application of the two previous lemmas.

Theorem V.5.1.

Let $\{P_n\}$ be a sequence of stochastic matrices such that $\lim_{n \to \infty} P_n = P$. Assume P is irreducible and periodic of period d. If $P^{d(i)}$ is weakly ergodic for $i = 0, 1, \ldots, d-1$, then $\{P_n\}$ is C-strongly ergodic. That is for all m,

$$\lim_{n \to \infty} \left\| \frac{1}{n} \sum_{t=1}^{n} P^{(m,m+t)} - Q \right\| = 0,$$

where Q is a constant stochastic matrix such that $QP = Q$.

Proof.   By Lemma V.5.2 we know that for all m,

$$\lim_{n \to \infty} \left\| \frac{1}{d} \sum_{l=0}^{d-1} P^{(m, m+n+l)} - Q \right\| = 0.$$

From this it follows by Lemma V.5.3 that for all m

$$\lim_{n \to \infty} \left\| \frac{1}{n} \sum_{t=1}^{n} P^{(m, m+t)} - Q \right\| = 0. \quad \blacktriangle$$

Remark V.5.1.   The assumption in the previous theorem and lemmas that P be irreducible was made for notational convenience. Actually it is only necessary that P corresponds to a Markov chain with one closed set of persistent states and possibly some transient states.

It is very hard to find simple applications of C-strong ergodicity. However, Theorem V.5.1 was motivated by a problem in the study of Markov decision processes. In particular, one is often interesred in finding an optimal policy that will minimize expected average cost and the Cesaro averages arise naturally in the calculation of average costs. For a detailed discussion of how Theorem V.5.1 is used in the study of Markov decision processes the reader is referred to Bowerman (1974).

There is yet another type of nonstationary Markov chain for which averages may converge even though the sequences themselves may not.

Example V.5.3.   Let $\{X_n\}$ be a Markov chain with transition matrices

$$P_{2n-1}\begin{pmatrix} 1 & 0 \\ \frac{1}{3} & \frac{2}{3} \end{pmatrix} \quad \text{and} \quad P_{2n} = \begin{pmatrix} 0 & 1 \\ 1 & 0 \end{pmatrix}$$

for $n = 1, 2, 3, \ldots$. This Markov chain is easily seen to be weakly ergodic but it is not strongly ergodic (Exercise 5). However, considering the cyclic pattern to the transition matrices one would expect the chain to be C-strongly ergodic. In fact, the subsequences $P_1 P_2 \cdots P_{2n-1} P_{2n}$ and $P_1 P_2 \cdots P_{2n} P_{2n+1}$ converge for this example (Exercise 5) so C-strong ergodicity follows from Lemma V.5.1. The reader must be careful, however, in deciding what the limiting matrix, Q, should be. From the theory given in Section V.4 one might conjecture that each row of Q is equal $\psi = \frac{1}{2}(\psi_1 + \psi_2)$ where $\psi_1 P_1 = \psi_1$ and $\psi_2 P_2 = \psi_2$. However, this is incorrect. In fact, $\psi = \frac{1}{2}(\psi_1^* + \psi_2^*)$ where $\psi_1^*(P_1P_2) = \psi_1^*$ and $\psi_2^*(P_2P_1) = \psi_2^*$. Hence for this example $Q = \begin{pmatrix} \frac{1}{2} & \frac{1}{2} \\ \frac{1}{2} & \frac{1}{2} \end{pmatrix}$ rather than $Q = \begin{pmatrix} \frac{3}{4} & \frac{1}{4} \\ \frac{3}{4} & \frac{1}{4} \end{pmatrix}$.

In view of Example V.5.3 we have the following lemma.

Lemma V.5.4. Let $\{X_n\}$ be a Markov chain with transition matrices $\{P_n\}$. Let $d > 1$ be fixed. Assume that $P_{nd+m} = P_m$ for $n = 0, 1, 2, \ldots$ and $m = 1, 2, \ldots, d$. If $P_1 P_2 \ldots P_d$ is a weakly ergodic stochastic matrix, then $\lim_{n\to\infty} P_1 P_2 \cdots P_{nd+m}$ exists in norm for $m = 1, 2, \ldots, d$. The limit is a constant matrix, Q_m with each row equal to the unique invariant probability vector of $(P_{(n-1)d+m+1} \cdots P_{nd+m})$.

Proof. Define $T_k = P_{kd+1} P_{kd+2} \cdots P_{kd+d}$ for $k = 0, 1, 2, \ldots$. The sequence of stochastic matrices, $\{T_k\}$ is stationary and weakly ergodic and hence strongly ergodic. Using an argument similar to that used in Theorem V.4.5 we can show that for $m = 1, 2, 3, \ldots, d$, $P_{kd+m+1} P_{kd+m+2} \cdots P_{kd+m+d}$ is also a weakly ergodic stochastic matrix. Let $m \in \{1, 2, \ldots, d\}$ be fixed and consider

$$P_1 P_2 \cdots P_{nd+m} = P_1 P_2 \cdots P_m (P_{m+1} \cdots P_{d+m})(P_{d+m+1} \cdots P_{2d+m})$$

$$\times (P_{2d+m+1} \cdots P_{3d+m}) \cdots (P_{(n-1)d+m+1} \cdots P_{nd+m}). \quad (5.7)$$

Let Q_m be a constant matrix with each of its rows equal to ψ_m where $\psi_m(P_{m+1} P_{m+2} \cdots P_{m+d}) = \psi_m$. It follows that

$$\|P_1 P_2 \cdots P_{nd+m} - Q_m\| = \|P_1 P_2 \cdots P_m P_{m+1} \cdots P_{nd+m} - P_1 P_2 \cdots P_m Q_m\|$$

$$\leqslant \|P_1 P_2 \cdots P_m\| \|(P_{m+1} P_{m+2} \cdots P_{m+d})^n - Q_m\|. \quad (5.8)$$

The right-hand side of (5.8) goes to zero as $n \to \infty$ since $(P_{m+1} \cdots P_{m+d})$ is strongly ergodic with limiting matrix Q_m. ▲

Using Lemmas V.5.1 and V.5.4 we get the following theorem.

Theorem V.5.2.

Let $\{X_n\}$ be a Markov chain with transition matrices $\{P_n\}$. Let $d > 1$ be fixed. Assume that $P_{nd+m} = P_m$ for $n = 0, 1, \ldots$ and $m = 1, 2, \ldots, d$. If $P_1 P_2 \ldots P_d$ is a weakly ergodic stochastic matrix, then $\{P_n\}$ is C-strongly ergodic.

Proof. By Lemma V.5.4 we have that $P_1 P_2 \ldots P_{nd+m}$ converges to a constant matrix, Q_m, as $n \to \infty$. Hence since these subsequences converge, Lemma V.5.1 gives convergence of the Cesaro averages to $Q = (Q_1 + Q_2 + \ldots + Q_d)/d$. ▲

Remark V.5.2. Theorem V.5.2 can be extended to the case where P_{nd+m} is not equal to P_m for all n but rather $\lim_{n \to \infty} P_{nd+m} = R_m$ and $R_1 R_2 \ldots R_m$ is weakly ergodic. The proof of this result is based on the results of Theorems V.4.5 and V.5.2. In fact, Theorems V.5.1 and V.5.2 can be extended to show that $\{P_n\}$ is uniformly C-strongly ergodic (i.e., the convergence holds uniformly in m). [See Bowerman, David, and Isaacson (1975).]

Example V.5.4. Let $\{X_n\}$ be a Markov chain with transition matrices defined to be

$$P_{3n-2} = \begin{bmatrix} \frac{1}{2} & \frac{1}{2} \\ 1 - \frac{1}{n} & \frac{1}{n} \end{bmatrix}, \qquad P_{3n-1} = \begin{bmatrix} 0 & 1 \\ 1 - \frac{1}{n} & \frac{1}{n} \end{bmatrix}$$

$$P_{3n} = \begin{bmatrix} \frac{1}{2} - \frac{1}{(n+1)} & \frac{1}{2} + \frac{1}{(n+1)} \\ \frac{1}{4} & \frac{3}{4} \end{bmatrix}$$

for $n = 1, 2, 3, \ldots$. It is easy to see that

$$\lim_{n \to \infty} P_{3n-2} = R_1 = \begin{pmatrix} \frac{1}{2} & \frac{1}{2} \\ 1 & 0 \end{pmatrix}, \qquad \lim_{n \to \infty} P_{3n-1} = R_2 = \begin{pmatrix} 0 & 1 \\ 1 & 0 \end{pmatrix},$$

and

$$\lim_{n \to \infty} P_{3n} = R_3 = \begin{pmatrix} \frac{1}{2} & \frac{1}{2} \\ \frac{1}{4} & \frac{3}{4} \end{pmatrix}.$$

The product of these limiting matrices is

$$R_1 R_2 R_3 = \begin{pmatrix} \frac{1}{2} & \frac{1}{2} \\ 1 & 0 \end{pmatrix} \begin{pmatrix} 0 & 1 \\ 1 & 0 \end{pmatrix} \begin{pmatrix} \frac{1}{2} & \frac{1}{2} \\ \frac{1}{4} & \frac{3}{4} \end{pmatrix} = \begin{pmatrix} \frac{3}{8} & \frac{5}{8} \\ \frac{1}{4} & \frac{3}{4} \end{pmatrix},$$

which is weakly ergodic. The invariant probability vector for $R_1 R_2 R_3$ is $\psi_0 = (\frac{2}{7}, \frac{5}{7})$, the invariant probability vector for $R_2 R_3 R_1$ is $\psi_1 = (\frac{6}{7}, \frac{1}{7})$, and the invariant probability vector for $R_3 R_1 R_2$ is $\psi_2 = (\frac{1}{7}, \frac{6}{7})$. Hence by Remark

V.5.2 we have

$$\lim_{n\to\infty} \frac{1}{n} \sum_{k=1}^{n} P_1 P_2 \cdots P_k = \frac{1}{3}\left[\begin{pmatrix} \frac{2}{7} & \frac{5}{7} \\ \frac{2}{7} & \frac{5}{7} \end{pmatrix} + \begin{pmatrix} \frac{6}{7} & \frac{1}{7} \\ \frac{6}{7} & \frac{1}{7} \end{pmatrix} + \begin{pmatrix} \frac{1}{7} & \frac{6}{7} \\ \frac{1}{7} & \frac{6}{7} \end{pmatrix} \right] = \begin{pmatrix} \frac{3}{7} & \frac{4}{7} \\ \frac{3}{7} & \frac{4}{7} \end{pmatrix}.$$

The interested reader might try to combine the results of Theorem V.5.1 and Theorem V.5.2 in order to show Cesaro strong ergodicity in the case where $\lim_{n\to\infty} P_{nd+m} = R_m$ and $R_1 R_2 \ldots R_d$ is irreducible and periodic of period d'.

This concludes the discussion of nonstationary Markov chains. In the next chapter we consider the problem of analyzing a Markov chain on a computer.

EXERCISES

1. Let $\{X_n\}$ be a Markov chain with

$$P_{4n-3} = \begin{pmatrix} 0 & 1 \\ 1 & 0 \end{pmatrix}, \qquad P_{4n-2} = \begin{pmatrix} \frac{1}{2} & \frac{1}{2} \\ \frac{1}{3} & \frac{2}{3} \end{pmatrix},$$

$$P_{4n-1} = \begin{pmatrix} 1 & 0 \\ 1 & 0 \end{pmatrix}, \qquad P_{4n} = \begin{pmatrix} 1 & 0 \\ 0 & 1 \end{pmatrix} \qquad \text{for } n = 1, 2, 3, \ldots$$

Find $\lim_{n\to\infty} (1/n) \sum_{k=1}^{n} P_1 P_2 \cdots P_k$.

2. Let $\{X_n\}$ be a Markov chain with

$$P_{3n-2} = \begin{pmatrix} \dfrac{1}{n} & 1 - \dfrac{1}{n} & 0 \\[2mm] \dfrac{1}{2} & \dfrac{1}{3} & \dfrac{1}{6} \\[2mm] \dfrac{1}{2} + \dfrac{1}{n+3} & \dfrac{1}{n+3} & \dfrac{1}{2} - \dfrac{2}{n+3} \end{pmatrix},$$

$$P_{3n-1} = \begin{pmatrix} 1 - \dfrac{2}{n+2} & \dfrac{1}{n+2} & \dfrac{1}{n+2} \\[2mm] \dfrac{1}{4} & \dfrac{1}{4} & \dfrac{1}{2} \\[2mm] \dfrac{1}{n+2} & \dfrac{1}{4} + \dfrac{1}{n+2} & \dfrac{3}{4} - \dfrac{2}{n+2} \end{pmatrix},$$

$$P_{3n} = \begin{bmatrix} \dfrac{1}{2} - \dfrac{1}{n+1} & \dfrac{1}{3} & \dfrac{1}{6} + \dfrac{1}{n+1} \\[2mm] \dfrac{1}{3} + \dfrac{1}{n+1} & \dfrac{1}{6} & \dfrac{1}{2} - \dfrac{1}{n+1} \\[2mm] \dfrac{1}{6} & \dfrac{1}{2} & \dfrac{1}{3} \end{bmatrix} \quad \text{for } n = 1, 2, 3, \ldots.$$

Find $\lim_{n \to \infty} (1/n) \sum_{k=1}^{n} P_1 P_2 \cdots P_k$.

3. Let $\{X_n\}$ be a Markov chain with

$$P_{3n-2} = \begin{bmatrix} 0 & 1 & 0 \\[2mm] 0 & \dfrac{1}{n} & 1 - \dfrac{1}{n} \\[2mm] 1 - \dfrac{1}{n} & 0 & \dfrac{1}{n} \end{bmatrix}, \quad P_{3n-1} = \begin{bmatrix} 0 & \dfrac{1}{2} & \dfrac{1}{2} \\[2mm] 1 - \dfrac{1}{n} & 0 & \dfrac{1}{n} \\[2mm] 1 - \dfrac{1}{n} & \dfrac{1}{n} & 0 \end{bmatrix},$$

$$P_{3n} = \begin{bmatrix} 1 - \dfrac{1}{n} & 0 & \dfrac{1}{n} \\[2mm] 0 & 1 - \dfrac{1}{n} & \dfrac{1}{n} \\[2mm] \dfrac{1}{n} & 0 & 1 - \dfrac{1}{n} \end{bmatrix} \quad \text{for } n = 1, 2, 3, \ldots.$$

Find $\lim_{n \to \infty} (1/n) \sum_{k=1}^{n} P_1 P_2 \cdots P_k$.

4. Let $\{X_n\}$ be a Markov chain with

$$P_n = \begin{bmatrix} \dfrac{1}{n+3} & \dfrac{1}{2} & \dfrac{1}{2} - \dfrac{2}{n+3} & \dfrac{1}{n+3} \\[2mm] \dfrac{1}{3} & 0 & \dfrac{1}{n+2} & \dfrac{2}{3} - \dfrac{1}{n+2} \\[2mm] 1 - \dfrac{2}{n+2} & 0 & \dfrac{1}{n+2} & \dfrac{1}{n+2} \\[2mm] 0 & 1 - \dfrac{1}{n+1} & 0 & \dfrac{1}{n+1} \end{bmatrix} \quad \text{for } n = 1, 2, 3, \ldots.$$

Find $\lim_{n \to \infty} (1/n) \sum_{k=1}^{n} P_1 P_2 \cdots P_k$.

5. Show that the Markov chain given in Example V.5.3 is not strongly ergodic. However, show that the subsequences $P_1 P_2 \cdots P_{2n}$ and $P_1 P_2 \cdots P_{2n} P_{2n+1}$ do converge.

6. Let P be an irreducible periodic stochastic matrix of period d. Let $P^{d(i)}$ represent the stochastic matrix operating on the ith cycle of P. (See Definition V.5.2.) Show that $P^{d(i)}$ is irreducible and aperiodic.

7. Assume $\lim_{n \to \infty} P^{nd} = Q_0$. Show that $\lim_{n \to \infty} P^{nd+l} = Q_0 P^l$.

8. Let $Q = (1/d) \sum_{l=0}^{d-1} Q_0 P^l$ where Q_0 and P are stochastic matrices. Show that $\| Q \| = 1$.

CHAPTER VI

Analysis of a Markov
Chain on a Computer

SECTION 1 **DETERMINING ERGODICITY**

The finite discrete-time stationary Markov chain has now been studied using three approaches. In all three approaches the transition matrix, P, was analyzed in order to determine whether or not the chain was ergodic. In the case of an ergodic chain, the long run distribution was calculated and if there were some transient states, the mean absorption times were found. In the case of a nonergodic chain, the chain was first written in block form and then all the states were classified according to persistency and periodicity. Finally the mean absorption times and absorption probabilities from the transient states were calculated. One of the prime objectives of this chapter is to consider how the above operations can be done on a computer when the transition matrix is finite. We will consider various alternative procedures that could be used and then recommend an algorithm that seems to be relatively efficient. It is hoped that by studying such an algorithm, the reader will improve his basic understanding of finite state space Markov chains. We also hope the reader will be challenged to try to write a better algorithm.

The first question one asks about a transition matrix is whether or not it is ergodic. The answering of this question will demand much attention in this chapter. The authors realize that in specific areas of applications, all the chains to be considered may be of the same type. Hence one would not be interested in having the computer waste time on the question of ergodicity. For example, in genetics the transition matrices typically have two absorbing states. Therefore by simply scanning the main diagonal of P for ones, it can be determined that the chain is nonergodic. However, the goal here is to write an algorithm that does not require any work by the user so it will admittedly be inefficient in certain specific cases.

Let P be an $n \times n$ stochastic matrix and consider the question of whether or not P is ergodic. There are many ways that this can be done. Using the theory of Chapter II one might calculate for each state, i, the set $C_i = \{ j : p_{ij}^{(n)} > 0 \text{ for some } n \}$. By comparing these it can be determined if P has exactly one irreducible closed subset and then the periodicity of the irreducible closed subsets of P can be found. Since this is not the procedure that will be suggested here, further details will not be given on this method. The reader might consider how the above analysis could be done efficiently on a computer. An efficient method for finding the closed sets of persistent states is given in Fox and Landi (1968).

Using the theory of Chapter IV the eigenvalues and eigenvectors of P could be calculated. If the eigenvalue 1 is strictly dominant, the corresponding left eigenvector is the long run distribution. If there are other eigenvalues of modulus one, they indicate either periodicity or reducibility as shown in Chapter IV. If one is only interested in the long run distribution when it exists, this approach might be useful. However, it does not classify the states of the chain, so mean absorption times and absorption probabilities could not be found with this method.

Finally the theory of Chapter V could and will be used to decide the question of ergodicity. Recall that a finite discrete-time stationary Markov chain is ergodic if and only if $\alpha(P^k) > 0$ for some k. Hence the general procedure for determining ergodicity will be to calculate the ergodic coefficient for high powers of P. In looking at the previous statement it becomes evident that the crucial question is how high the power of P must be. Obviously the algorithm cannot tell the computer to take higher and higher powers of P until $\alpha(P^k) > 0$ since for nonergodic matrices this would never stop. Therefore, in order to be able to use the computer for determining whether $\alpha(P^k)$ will be greater than zero for some k, we must be able to answer the following question. If P is an $n \times n$ stochastic matrix, does there exist a $k_0 = k_0(n)$ such that $\alpha(P^k) > 0$ for some k if and only if $\alpha(P^{k_0}) > 0$? [An alternative way of expressing this question is to ask whether the property of $\alpha(P^k)$ being greater than zero for some k is "decidable."]

Paz (1963) proved that this property for stochastic matrices is indeed decidable by showing that for an $n \times n$ stochastic matrix, $\alpha(P^k) > 0$ for some k if and only if $\alpha(P^{n(n-1)/2}) > 0$. It was not known at that time whether or not some power of P smaller than $n(n-1)/2$ would suffice. It was later shown by Madsen (1975) that $k_0(n) = [(n-1)^2/2 + 1]$, where $[\cdot]$ represents the greatest integer function, is the smallest such power of P that can be used for determining the ergodicity of P (Exercise 2). The proof given by Madsen will not be included here because of its length and because it uses concepts from number theory with which many readers might not be familiar. However, we will give a proof of the slightly weaker result given by Paz. We include this proof since it will allow us to introduce some definitions and concepts that will be useful in discussions given later in this chapter. Also the proof may give some further insight into the relationship between the ergodic coefficient and ergodicity.

Definition VI.1.1. *Let $\{X_n\}$ be a stationary finite Markov chain with transition matrix P. States i and j are said to have state k as a common consequent of order m if $p_{ik}^{(m)} > 0$ and $p_{jk}^{(m)} > 0$.*

The proof of Paz's result follows immediately from the next two lemmas.

Lemma VI.1.1. *If P is a finite stochastic matrix, then $\alpha(P^m) > 0$ if and only if every pair of states has a common consequent of order m.*

Proof. The proof is left to the reader (Exercise 7). ▲

Lemma VI.1.2. *Let P be an $n \times n$ stochastic matrix. If states i and j have a common consequent of some order, then they must have a common consequent of order $n(n-1)/2$.*

Proof. It suffices to show that states i and j have a common consequent of order less than or equal to $n(n-1)/2$, since if a pair of states has a common consequent of order m, then they have a common consequent of all higher orders (Exercise 3). So assume that states i and j have state k as a common consequent of order N and that they have no state which is a common consequent of order less than N. That is, there exist paths

$$i = i_0 \to i_1 \to i_2 \to \cdots \to i_{N-1} \to k,$$

$$j = j_0 \to j_1 \to j_2 \to \cdots \to j_{N-1} \to k$$

such that $p_{ii_1} p_{i_1 i_2} \cdots p_{i_{N-1} k} > 0$ and $p_{jj_1} p_{j_1 j_2} \cdots p_{j_{N-1} k} > 0$. For notational convenience, denote the position of these two paths at time m by the ordered pair (i_m, j_m). We first note that no pair (i_m, j_m), $m < N$, can have its elements equal, for if $i_m = j_m$, then $i = i_0 \to i_1 \to \cdots \to i_m$ and $j = j_0 \to j_1 \to \cdots \to j_m = i_m$ are paths of positive probability (why?) of length $m < N$. This would imply that state i_m is a common consequent of states i and j of order $m < N$, which contradicts the assumption that there is no state which is a common consequent of states i and j of order less than N.

We next note that no two pairs can be equal, for if they were, this would imply that states i and j would have state k as a common consequent of order less than N, contradicting our assumption. To see this, note that if

$m_1 < m_2$ and if $(i_{m_1}, j_{m_1}) = (i_{m_2}, j_{m_2})$, then $i = i_0 \to i_1 \to \cdots i_{m_1} = i_{m_2} \to i_{m_2+1} \to \cdots$ $\to k$ and $j = j_0 \to j_1 \to \cdots \to j_{m_1} = j_{m_2} \to j_{m_2+1} \to \cdots \to k$ are both paths of positive probability of length $N - (m_2 - m_1)$.

Finally we point out that a similar contradiction would be reached if two pairs have the same elements, but in reverse order. That is if $m_1 < m_2$ and if (i_{m_1}, j_{m_1}) and (i_{m_2}, j_{m_2}) satisfy $i_{m_1} = j_{m_2}$ and $j_{m_1} = i_{m_2}$, then

$$i = i_0 \to i_1 \to \cdots \to i_{m_1} = j_{m_2} \to j_{m_2+1} \to \cdots \to j_{N-1} \to k$$

and

$$j = j_0 \to j_1 \to \cdots \to j_{m_1} = i_{m_2} \to i_{m_2+1} \to \cdots \to i_{N-1} \to k$$

are both paths of positive probability of length $N - (m_2 - m_1)$.

In view of the restrictions that no pair can have its elements equal and that no two pairs can be equal (without regard to order) we know that the set of pairs $(i_m, j_m), m = 0, 1, \ldots, N-1$ can contain at most $\binom{n}{2}$ pairs (since $\binom{n}{2}$ is the number of distinct unordered pairs of numbers that can be drawn, without repetition, from a set of n numbers). Hence $N \leqslant \binom{n}{2}$ $= n(n-1)/2$. ▲

Theorem VI.1.1.

Let P be the $n \times n$ stochastic transition matrix for a stationary Markov chain. $\alpha(P^k) > 0$ for some k if and only if $\alpha(P^{n(n-1)/2}) > 0$.

Proof. The proof follows immediately from Lemmas VI.1.1 and VI.1.2. ▲

Now that we have shown that the question of ergodicity for an $n \times n$ stochastic matrix is decidable on a computer, we consider how this decision can be reached efficiently. We first note that the numerical values of the probabilities of transitions are not needed to decide this question, but rather only whether or not the probability is nonzero. Therefore the first step is to transform the stochastic matrix, P, into an adjacency matrix, Z, consisting of zeros and ones. (See, for example, Anderson 1970.) That is, replace the positive entries of P by ones. The matrix Z is not stochastic but we can still define the ergodic coefficient of Z to be

$$\alpha(Z) = \min_{i,k} \sum_j \min(z_{ij}, z_{kj}).$$

The important fact to note is that $\alpha(Z^k)$ will be positive if and only if $\alpha(P^k)$ is positive so Paz's criterion for deciding ergodicity can be applied to Z. There is a possibility of creating some large numbers as entries in Z^k for large k. To avoid this situation, the positive integers in $Z^2, Z^3, \ldots, Z^k$ are again reduced to ones so by Z^k we will actually mean the adjacency matrix of Z^k (i.e., $z_{ij}^{(k)} = \min\{1, \sum_{l \in S} z_{il}^{(k-1)} z_{lj}\}$). It is again easy to see that $\alpha(Z^k) > 0$ if and only if $\alpha(P^k) > 0$.

Example VI.1.1. Consider the stochastic matrix

$$P = \begin{bmatrix} 0 & 0 & 1 \\ 0 & \frac{1}{3} & \frac{2}{3} \\ \frac{1}{4} & \frac{3}{4} & 0 \end{bmatrix}.$$

Then

$$Z = \begin{bmatrix} 0 & 0 & 1 \\ 0 & 1 & 1 \\ 1 & 1 & 0 \end{bmatrix} \quad \text{and} \quad Z^2 = \begin{bmatrix} 1 & 1 & 0 \\ 1 & 1 & 1 \\ 0 & 1 & 1 \end{bmatrix}.$$

All higher powers of Z contain all ones. Note that $\alpha(P) = \alpha(Z) = 0$ and $\alpha(P^2) > 0$ and $\alpha(Z^2) > 0$.

One advantage of using the adjacency matrices rather than the actual P matrices is that all of the matrix multiplications are done with matrices of zeros and ones, so fixed point arithmetic can be used and all the computations are numerically stable. Another advantage of using a zero–one matrix for Z^k is that

$$\sum_{l=1}^{n} \min(z_{il}, z_{jl}) = \sum_{l=1}^{n} z_{il} z_{jl}$$

and it is computationally faster to calculate the products than make the necessary comparisons.

It can be shown that if $\alpha(Z^k) > 0$, then $\alpha(Z^m) > 0$ for $m > k$ (Exercise 4). Hence it is more efficient to calculate the powers $Z, Z^2, Z^4, Z^8, \ldots, Z^{2^k}$ than $Z, Z^2, Z^3, \ldots, Z^{n(n-1)/2}$. Using the first sequence the ergodicity question is decided when $2^k \geqslant n(n-1)/2$ and for moderately large n the reduction in steps is significant. (Note that we could use $[(n-1)^2/2 + 1]$ in place of $n(n-1)/2$ here, but we have chosen to use the latter value since it was obtained from a theorem proven in this book.)

With the above adjacency matrices, Z^{2^k}, in mind, we return to Paz's criterion for determining ergodicity. Two obvious algorithms are suggested immediately. One could find Z^{2^k} for $2^k \geqslant n(n-1)/2$ and then calculate $\alpha(Z^{2^k})$ or one could calculate $\alpha(Z^{2^k})$ for each k until $\alpha(Z^{2^k}) > 0$ or $2^k \geqslant n(n-1)/2$. Clearly the first procedure is inefficient if the chain is ergodic and the second procedure is inefficient if the chain is nonergodic.

Example VI.1.2. Let $\{X_n\}$ be a discrete-time stationary Markov chain with transition matrix

$$P = \begin{bmatrix} 0 & 0 & 1 & 0 \\ 0 & 0 & 1 & 0 \\ \frac{1}{3} & 0 & \frac{2}{3} & 0 \\ \frac{1}{2} & \frac{1}{2} & 0 & 0 \end{bmatrix}.$$

For this chain, $\alpha(Z^2) > 0$ so the ergodicity can be determined at step two. However if one decides to calculate Z^{2^k} until $2^k \geqslant n(n-1)/2 = 6$ then Z, Z^2, Z^4 and Z^8 would be calculated before discovering $\alpha(Z^8) > 0$. On the other hand if the transition matrix had the form

$$P^* = \begin{bmatrix} 0 & 0 & 1 & 0 \\ 0 & 0 & 1 & 0 \\ \frac{1}{3} & \frac{2}{3} & 0 & 0 \\ \frac{1}{2} & \frac{1}{2} & 0 & 0 \end{bmatrix},$$

then the corresponding Markov chain would have a periodic subclass so $\alpha(Z^k)$ would be zero for all k. In this case it would be inefficient to calculate $\alpha(Z^{2^k})$ at each step since the ergodic coefficient would be zero and the higher powers of Z^{2^k} would have to be calculated anyway.

As a compromise between these two extremes one might use some property of Z^{2^k} to decide whether the ergodic coefficient should be calculated. Intuitively, the ergodic coefficient should be calculated if there is a good chance that it will be positive. If it looks as though the ergodic coefficient will be zero, the matrix Z^{2^k} should be squared. As an attempt to formalize this intuitive procedure we describe the "column sum method" for determining ergodicity.

At each step the columns of Z^{2^k} are summed. If any column sums to n, then the ergodic coefficient is positive, so the chain must be ergodic (Exercise 5). If any column sums to $n-1$, there is a reasonable chance that the ergodic coefficient will be positive. In any event the calculation of the ergodic coefficient in this case is greatly simplified. If the jth column of Z^{2^k} has only one zero and this zero appears in row i, then in checking for positivity of $\alpha(Z^{2^k})$ one need only compare row i with the other $n-1$ rows.

In general there are $\binom{n}{2}$ such comparisons to make, so this is a substantial savings. The "column sum method" reduces to the following rules.

1. At each step sum the columns of Z^{2^k}.
2. If any column of Z^{2^k} sums to n, stop, since the chain is ergodic.
3. If any column of Z^{2^k} sums to $n-1$, calculate $\alpha(Z^{2^k})$. If $\alpha(Z^{2^k})>0$ stop, since the chain is ergodic. If $\alpha(Z^{2^k})=0$, square the matrix Z^{2^k} and repeat the above.
4. If no columns sum to $n-1$ or n, square Z^{2^k} and repeat the above.

These rules are summarized in the flow chart given in Figure VI.1.1.

The following example illustrates the column sum method for determining ergodicity.

Example VI.1.3. Using the matrices of Example VI.1.2 we see for the first matrix that

$$
Z = \begin{bmatrix} 0 & 0 & 1 & 0 \\ 0 & 0 & 1 & 0 \\ 1 & 0 & 1 & 0 \\ 1 & 1 & 0 & 0 \end{bmatrix}
$$

so the third column sums to $n-1$. However, $\alpha(Z)=0$ so we calculate

$$
Z^2 = \begin{bmatrix} 1 & 0 & 1 & 0 \\ 1 & 0 & 1 & 0 \\ 1 & 0 & 1 & 0 \\ 0 & 0 & 1 & 0 \end{bmatrix}.
$$

The third column of Z^2 sums to n so we know P is ergodic.
For the second matrix we have

$$
Z^* = \begin{bmatrix} 0 & 0 & 1 & 0 \\ 0 & 0 & 1 & 0 \\ 1 & 1 & 0 & 0 \\ 1 & 1 & 0 & 0 \end{bmatrix}.
$$

Since the largest column sum for Z^* is two we go directly to

$$
(Z^*)^2 = \begin{bmatrix} 1 & 1 & 0 & 0 \\ 1 & 1 & 0 & 0 \\ 0 & 0 & 1 & 0 \\ 0 & 0 & 1 & 0 \end{bmatrix}
$$

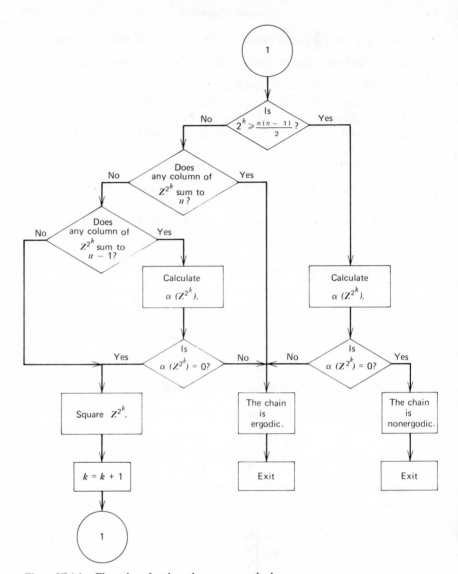

Figure VI.1.1. Flow chart for the column sum method.

which again has a largest column sum of 2. We continue to $(Z^*)^8$ without ever calculating an ergodic coefficient. Since $\alpha[(Z^*)^8] = 0$, Z^* is nonergodic.

Many other criteria could be used at each step to decide if the ergodic coefficient should be calculated. However, since the algorithm to be given here will use the column sum method, the alternatives will not be discussed. The interested reader might consider some alternatives and compare their strengths and weaknesses with the column sum method.

EXERCISES

1. Given the following stochastic matrices, list all distinct pairs of states and, for each pair, list the common consequences of order $m = 1$, 2, and 3. In view of Lemmas VI.1.1 and VI.1.2, what conclusions can be drawn?

$$\text{(a)} \quad P = \begin{bmatrix} 1 & 0 & 0 \\ 0 & 1 & 0 \\ 0 & \frac{2}{3} & \frac{1}{3} \end{bmatrix}, \qquad \text{(b)} \quad P = \begin{bmatrix} \frac{1}{2} & 0 & \frac{1}{2} \\ 1 & 0 & 0 \\ 0 & \frac{2}{3} & \frac{1}{3} \end{bmatrix}$$

$$\text{(c)} \quad P = \begin{bmatrix} 0 & 0 & 1 \\ 0 & 0 & 1 \\ \frac{1}{3} & \frac{2}{3} & 0 \end{bmatrix}$$

†2. Construct an example of a 4×4 stochastic matrix, for which $\alpha(P^{[(n-1)^2/2+1]}) > 0$ but $\alpha(P^{[(n-1)^2/2]}) = 0$. This shows that the bound given by Madsen (1975) is as small as possible for $n = 4$.

3. Show that if states i and j have a common consequent of order m, they have a common consequent of order $m + 1$.

*4. Show that if $\alpha(P^k) > 0$ and if $k \leqslant m$, then $\alpha(P^m) > 0$.

*5. Show that if P is finite and if any column of P has no zeros, then $\alpha(P) > 0$. (In particular, show that this implies that any state corresponding to a column with no zeros must be aperiodic.)

6. Describe a procedure for deciding whether or not the ergodic coefficient of Z^{2^k} should be calculated at step k. Compare this method with the column sum method.

7. Show that if P is a finite stochastic matrix, then $\alpha(P^m) > 0$ if and only if every pair of states has a common consequent of order m.

Are either of the above implications true for infinite stochastic matrices? Does the result hold if P is non-negative but not necessarily stochastic?

SECTION 2   ANALYSIS OF THE CHAIN

At this point the stochastic matrix has been determined to be either ergodic or nonergodic. The next step is to analyze the chain.

Most of the results of this section are proved by using the following simple lemma.

Lemma VI.2.1.   Let P be an $n \times n$ stochastic matrix. If $i \neq j$ and $p_{ij}^{(k)} > 0$ for some k, then $p_{ij}^{(k)} > 0$ for some $k \leqslant n - 1$. Also if $p_{ii}^{(k)} > 0$ for some $k > 0$, then $p_{ii}^{(k)} > 0$ for some $k \leqslant n$.

Proof.   Assume $p_{ij}^{(k)} > 0$ for some k. This means there must be at least one path from i to j which has positive probability. That is $p_{ij}^{(k)} \geqslant p_{ii_1} p_{i_1 i_2} \cdots p_{i_{k-1} j} > 0$. If all the loops are eliminated from this path, there remains a path with positive probability from i to j with $n - 1$ steps or less. For example assume $X_0 = i_0$, $X_1 = i_1, \ldots X_{k-1} = i_{k-1}$ and $X_k = j$. Now if $i_1 = i_8$, say, we can consider a new path $X_0 = i_0, X_1 = i_1, X_2 = i_9, X_3 = i_{10}, \ldots X_{k-7} = j$. Note that $p_{ii_1} p_{i_1 i_9} \cdots p_{i_{k-1} j} \geqslant p_{ii_1} p_{i_1 i_2} \cdots p_{i_{k-1} j} \geqslant 0$ so the new path has positive probability. By eliminating loops in this manner we get a path with positive probability such that no state is visited more than once. Since there are only $n - 1$ states distinct from i, state j must be visited in $n - 1$ steps or less. The case $i = j$ is proved in a similar fashion.   ▲

We now consider the problem of analyzing a stochastic matrix that is nonergodic. The first goal is to write the matrix in block form. The basic idea behind this construction is to start with a persistent state and generate the irreducible closed subset containing this state. Continue this until there are no more persistent states. Hence we see that the first problem is to find a persistent state.

By definition state i is persistent if

$$\sum_{n=0}^{\infty} p_{ii}^{(n)} = \infty \quad \text{or} \quad \sum_{n=0}^{\infty} f_{ii}^{(n)} = 1.$$

These conditions would be impossible to check on a computer in most cases. Therefore some alternative method for finding a persistent state in a nonergodic stochastic matrix must be given. In order to develop such a method we first must revise the column sum method for determining ergodicity. In particular the original column sum method required that powers Z^{2^k} be taken until $\alpha(Z^{2^k}) > 0$ or $2^k \geqslant n(n-1)/2$. We now require that powers be taken until $\alpha(Z^{2^k}) > 0$ or $2^k \geqslant (n-1)^2$. The significance of the power $(n-1)^2$ will become apparent in the next theorem. For notational convenience let M be the power of Z at which the ergodicity question is decided. For nonergodic chains we now have $M \geqslant (n-1)^2$, so the following theorem can be used to find a persistent state. [The reader may find the proof of this theorem to be somewhat difficult to follow. It may be helpful to go through the proof in the special case where m (as defined in the proof) equals one and where there are only one or two transient states.]

Theorem VI.2.1.

Let P be a nonergodic stochastic matrix. State i is persistent if the sum of the elements in the ith column of the adjacency matrix, Z^M, is a maximum. (That is, $\sum_{j=1}^{n} z_{ji} = \max_k \sum_{j=1}^{n} z_{jk}$.)

Proof. Let $T \neq \varnothing$ denote the set of transient states and $C_1, C_2, \ldots, C_m$ denote the irreducible closed sets of persistent states. (If $T = \varnothing$ then all states are persistent so the state with largest column sum is persistent.) For the sake of argument assume the block forms of Z and Z^M are

$$
Z = \begin{bmatrix}
Z_1 & 0 & \cdots & 0 & 0 \\
0 & Z_2 & \cdots & 0 & 0 \\
\cdot & \cdot & \cdots & \cdot & \cdot \\
\cdot & \cdot & \cdots & \cdot & \cdot \\
0 & 0 & \cdots & Z_m & 0 \\
A_1 & A_2 & \cdots & A_m & B
\end{bmatrix}
$$

and

$$
Z^M = \begin{bmatrix}
Z_1^M & 0 & \cdots & 0 & 0 \\
0 & Z_2^M & \cdots & 0 & 0 \\
\cdot & \cdot & \cdots & \cdot & \cdot \\
\cdot & \cdot & \cdots & \cdot & \cdot \\
0 & 0 & \cdots & Z_m^M & 0 \\
A_1^* & A_2^* & \cdots & A_m^* & B^M
\end{bmatrix}
$$

where Z_i is the adjacency matrix for C_i and B is the adjacency matrix for T. (Note that the matrix Z^M in the computer is not yet in this block form but it is helpful for this proof to picture it this way.) Let B^M contain a column of l ones. It is sufficient to show the existence of a column of Z^M corresponding to the states in $\bigcup_{i=1}^{m} C_i$ with at least $l+1$ ones. Without loss of generality assume column $j \in B^M$ has l ones. This says state j can be reached in M steps from l transient states, $\{t_1, t_2, \ldots, t_l\}$. From j it is possible to reach at least one persistent state. Again without loss of generality assume $Z_{ji}^{(N)} = 1$ for some N and $i \in C_1$. Hence using the transition matrix, Z^{M+N} it is possible to go from these l transient states to state i. Under the assumption that there are some transient states the dimension of C_1 is at most $n-1$. Hence by Lemma VI.2.1 there exists an integer $b \leqslant n-1$ such that $z_{ii}^{(b)} = 1$. Consider the two cases, C_1 aperiodic and C_1 periodic, separately.

1. If C_1 is aperiodic, then it is possible to go from these l states in T to state i using the transition matrices $Z^b, Z^{2b}, \ldots$. That is, there is no danger of the chain with this transition matrix continuing to avoid state i since C_1 is aperiodic (Exercise 6). Since there are only n states, all transitions that are possible can be done in $n-1$ steps or less. Hence it is possible to go from each of these l states to state i in $n-1$ steps or less using the transition matrix Z^b. Note that under the transition matrix Z^b once the process gets to state i, it stays with positive probability since $z_{ii}^{(b)} > 0$. Hence $Z^{b(n-1)}$ has $z_{t_j i}^{b(n-1)} = 1$ for these l transient states, $\{t_1, t_2, \ldots, t_l\}$. Also $z_{ik}^{[M-b(n-1)]} = 1$ for some $k \in C_1$ since C_1 is closed, so

$$1 \geqslant z_{t_j k}^{(M)} \geqslant z_{t_j i}^{b(n-1)} \cdot z_{ik}^{[M-b(n-1)]} = 1$$

for these l transient states, $\{t_1, t_2, \ldots, t_l\}$. The matrix corresponding to the set of persistent states, C_1, has at least one one in each column (Exercise 4). Hence the kth column of Z^M has at least $l+1$ ones.

2. Let C_1 have period d. Then C_1 is partitioned into d cyclic classes, $D_1, D_2, \ldots, D_d$, such that under the transition matrix, P, the chain moves from D_j to D_{j+1} and from D_d to D_1. Each of these cyclic classes is the state space for an irreducible aperiodic Markov chain under the transition matrix Z^d. Hence this case is reduced to case 1 by using Z^d as the transition matrix instead of Z and using one of the D_k's instead of C_1. Note that since it is possible to go from the l transient states under consideration to state j in M steps, it is possible to go from these l states into at least one of the D_k's in a number of steps which is a multiple of d. For the sake of argument assume that $Z_{t_j i}^{md} = 1$ for the l transient states

$\{t_i, t_2, \ldots, t_l\}$. Also note that the integer b, chosen in Case 1 so that $z_{ii}^{(b)} = 1$, will now be of the form $b = kd \leqslant n - 1$. The details of this case are left as an exercise. ▲

Example VI.2.1. Let $\{X_n\}$ be a Markov chain with transition matrix

$$P = \begin{pmatrix} \frac{1}{2} & 0 & 0 & \frac{1}{2} & 0 & 0 \\ 0 & \frac{1}{3} & \frac{2}{3} & 0 & 0 & 0 \\ 0 & \frac{1}{2} & \frac{1}{8} & \frac{1}{8} & 0 & \frac{1}{4} \\ \frac{1}{3} & 0 & 0 & \frac{2}{3} & 0 & 0 \\ 0 & 0 & 0 & 0 & 1 & 0 \\ 0 & \frac{1}{4} & \frac{1}{2} & 0 & 0 & \frac{1}{4} \end{pmatrix}.$$

By the methods of Chapter II we see that this chain has two closed sets, $C_1 = \{1, 4\}$ and $C_2 = \{5\}$, and the remaining states are transient. Using the methods of this chapter we get

$$Z = \begin{pmatrix} 1 & 0 & 0 & 1 & 0 & 0 \\ 0 & 1 & 1 & 0 & 0 & 0 \\ 0 & 1 & 1 & 1 & 0 & 1 \\ 1 & 0 & 0 & 1 & 0 & 0 \\ 0 & 0 & 0 & 0 & 1 & 0 \\ 0 & 1 & 1 & 0 & 0 & 1 \end{pmatrix}$$

and

$$Z^{32} = \begin{pmatrix} 1 & 0 & 0 & 1 & 0 & 0 \\ 1 & 1 & 1 & 1 & 0 & 1 \\ 1 & 1 & 1 & 1 & 0 & 1 \\ 1 & 0 & 0 & 1 & 0 & 0 \\ 0 & 0 & 0 & 0 & 1 & 0 \\ 1 & 1 & 1 & 1 & 0 & 1 \end{pmatrix}.$$

Now $32 > (n-1)^2 = 25$ so columns 1 and 4 correspond to persistent states. Thus Theorem III.2.1 does provide a method of finding a persistent state. However, if one is interested in finding all the persistent states, more theory is needed.

Once a persistent state, i, is found the irreducible closed set, C_i, generated by this state is constructed as follows. Let $K_0 = \{j : z_{ij}^{(M)} = 1\}$. If state i is aperiodic, K_0 is the complete irreducible closed set generated by state i (Exercise 9). However, since i may be periodic, we let K_1

$= \{ j : z_{ij}^{(M+1)} = 1 \}$. If $K_1 = K_0$, then we know state i is aperiodic and $K_0 = C_i$. If $K_1 \neq K_0$, then we continue to find $K_2 = \{ j : z_{ij}^{(M+2)} = 1 \}$ and so on until $K_m = K_0$. The smallest integer d for which $K_d = K_0$ is the period of C_i. The set C_i is defined to be $C_i = \bigcup_{j=0}^{d-1} K_j$. This process yields the irreducible closed set generated by state i and defines its period.

The next step is to find another persistent state and its associated closed set. It would certainly be nice if one could go directly to the column with the next largest number of ones. Returning to Example VI.2.1 we see that this approach will not help us to find another persistent state. In this example state 1 is persistent and $C_1 = \{1,4\}$. The columns of Z^M with the next largest number of ones are columns 2, 3, and 6. Unfortunately all these columns correspond to transient states.

We can, however, find another persistent state by using the following procedure. At this point all of the states in C_i have been classified as persistent states and hence may be omitted from further consideration. Furthermore, states outside of C_i from which C_i can be reached must be transient, hence they may also be classified at this time.

In view of these remarks the matrix Z^M should be changed as follows. Place zeros in all rows and columns corresponding to the states in C_i. Also place zeros in all rows and columns corresponding to states from which C_i can be reached. The following steps will allow the computer to perform the above operations. First choose one state k_j from each of the disjoint sets K_j in C_i. Define

$$D_0 = \left\{ j : z_{j,k_0}^{(M)} = 1 \right\}, \qquad D_1 = \left\{ j : z_{j,k_1}^{(M)} = 1 \right\}, \dots \quad \text{and} \quad D_{d-1} = \left\{ j : z_{j,k_{d-1}}^{(M)} = 1 \right\}.$$

Let $D = \bigcup_{l=0}^{d-1} D_l$. The set D includes all the persistent states that intercommunicate with state i and all the transient states from which state i can be reached. (Why?) Now define a new matrix Z^* such that $z_{ij}^* = z_{ij}^{(M)}$ if both i and j belong to the complement of D and let $z_{ij}^* = 0$ otherwise. Any column of Z^* with a maximum sum again corresponds to a persistent state so the procedure described earlier for finding its associated irreducible closed set can be used (Exercise 8). Returning to Example VI.2.1 we see that $C_1 = \{1,4\}$ but $D = \{1,2,3,4,6\}$ so

$$Z^* = \begin{bmatrix} 0 & 0 & 0 & 0 & 0 & 0 \\ 0 & 0 & 0 & 0 & 0 & 0 \\ 0 & 0 & 0 & 0 & 0 & 0 \\ 0 & 0 & 0 & 0 & 0 & 0 \\ 0 & 0 & 0 & 0 & 1 & 0 \\ 0 & 0 & 0 & 0 & 0 & 0 \end{bmatrix}.$$

The column with the largest number of ones is column five so state 5 is persistent. It follows immediately that $K_0 = K_1 = \{5\}$ so the closed set associated with state 5 is the absorbing set $\{5\}$.

In general, the reduction of Z^M continues until the matrix contains only zeros. At this point all the irreducible closed subsets and their periods have been identified. The remaining states are transient. At this point it is easy to rewrite P into block form since this simply involves an interchange of the appropriate rows and columns of P.

The final step in the analysis of a nonergodic stochastic matrix is to find the mean absorption times and the absorption probabilities from the transient states. This is straightforward once the matrix has been written in block form. As derived in Chapter III one simply calculates $N = (I - Q)^{-1}$ and then looks at $N\mathbf{1}'$ and NR where $\mathbf{1}$ and R are as defined in Chapter III. Note that the problem of inverting a matrix on a computer has been studied by many authors. (Fadeev and Fadeeva, 1963, and Westlake, 1968, are just two of many possible references.) Hence we will not dwell on this part of the problem here.

Example VI.2.2. Let $\{X_n\}$ be a discrete-time stationary Markov chain with transition matrix

$$P = \begin{bmatrix} 0 & 0 & 0 & 0 & 1 & 0 & 0 & 0 & 0 & 0 \\ 0 & \frac{1}{2} & 0 & \frac{1}{2} & 0 & 0 & 0 & 0 & 0 & 0 \\ 0 & 0 & \frac{1}{2} & 0 & \frac{1}{2} & 0 & 0 & 0 & 0 & 0 \\ 0 & 0 & 0 & 0 & 0 & 0 & 0 & 1 & 0 & 0 \\ 1 & 0 & 0 & 0 & 0 & 0 & 0 & 0 & 0 & 0 \\ 0 & 0 & 0 & 0 & 0 & 1 & 0 & 0 & 0 & 0 \\ \frac{1}{4} & 0 & \frac{1}{4} & 0 & \frac{1}{4} & \frac{1}{4} & 0 & 0 & 0 & 0 \\ 0 & 0 & 0 & 0 & 0 & 0 & 0 & 0 & 1 & 0 \\ 0 & 1 & 0 & 0 & 0 & 0 & 0 & 0 & 0 & 0 \\ \frac{1}{10} & \frac{1}{10} & \frac{1}{10} & \frac{1}{10} & \frac{1}{10} & \frac{1}{10} & \frac{1}{10} & \frac{1}{10} & \frac{1}{10} & \frac{1}{10} \end{bmatrix}.$$

The corresponding Z matrix is

$$Z = \begin{bmatrix} 0 & 0 & 0 & 0 & 1 & 0 & 0 & 0 & 0 & 0 \\ 0 & 1 & 0 & 1 & 0 & 0 & 0 & 0 & 0 & 0 \\ 0 & 0 & 1 & 0 & 1 & 0 & 0 & 0 & 0 & 0 \\ 0 & 0 & 0 & 0 & 0 & 0 & 0 & 1 & 0 & 0 \\ 1 & 0 & 0 & 0 & 0 & 0 & 0 & 0 & 0 & 0 \\ 0 & 0 & 0 & 0 & 0 & 1 & 0 & 0 & 0 & 0 \\ 1 & 0 & 1 & 0 & 1 & 1 & 0 & 0 & 0 & 0 \\ 0 & 0 & 0 & 0 & 0 & 0 & 0 & 0 & 1 & 0 \\ 0 & 1 & 0 & 0 & 0 & 0 & 0 & 0 & 0 & 0 \\ 1 & 1 & 1 & 1 & 1 & 1 & 1 & 1 & 1 & 1 \end{bmatrix}.$$

Since the chain is nonergodic and since $n = 10$, we must take $k = 7$ so $2^k = 128 \geqslant 81$.

$$Z^{128} = \begin{bmatrix} 1 & 0 & 0 & 0 & 0 & 0 & 0 & 0 & 0 & 0 \\ 0 & 1 & 0 & 1 & 0 & 0 & 0 & 1 & 1 & 0 \\ 1 & 0 & 1 & 0 & 1 & 0 & 0 & 0 & 0 & 0 \\ 0 & 1 & 0 & 1 & 0 & 0 & 0 & 1 & 1 & 0 \\ 0 & 0 & 0 & 0 & 1 & 0 & 0 & 0 & 0 & 0 \\ 0 & 0 & 0 & 0 & 0 & 1 & 0 & 0 & 0 & 0 \\ 1 & 0 & 1 & 0 & 1 & 1 & 0 & 0 & 0 & 0 \\ 0 & 1 & 0 & 1 & 0 & 0 & 0 & 1 & 1 & 0 \\ 0 & 1 & 0 & 1 & 0 & 0 & 0 & 1 & 1 & 0 \\ 1 & 1 & 1 & 1 & 1 & 1 & 1 & 1 & 1 & 1 \end{bmatrix}.$$

Hence states 2, 4, 8, and 9 are persistent. Starting with state 2 we get $K_0 = \{2,4,8,9\}$ and $K_1 = \{2,4,8,9\}$ so the irreducible closed set $\{2,4,8,9\}$ is aperiodic. The set D corresponding to this closed set is $D = D_0 = \{2,4,8,9,10\}$ so the new matrix Z^* is

$$Z^* = \begin{bmatrix} 1 & 0 & 0 & 0 & 0 & 0 & 0 & 0 & 0 & 0 \\ 0 & 0 & 0 & 0 & 0 & 0 & 0 & 0 & 0 & 0 \\ 1 & 0 & 1 & 0 & 1 & 0 & 0 & 0 & 0 & 0 \\ 0 & 0 & 0 & 0 & 0 & 0 & 0 & 0 & 0 & 0 \\ 0 & 0 & 0 & 0 & 1 & 0 & 0 & 0 & 0 & 0 \\ 0 & 0 & 0 & 0 & 0 & 1 & 0 & 0 & 0 & 0 \\ 1 & 0 & 1 & 0 & 1 & 1 & 0 & 0 & 0 & 0 \\ 0 & 0 & 0 & 0 & 0 & 0 & 0 & 0 & 0 & 0 \\ 0 & 0 & 0 & 0 & 0 & 0 & 0 & 0 & 0 & 0 \\ 0 & 0 & 0 & 0 & 0 & 0 & 0 & 0 & 0 & 0 \end{bmatrix}.$$

We see that the states 1 and 5 are persistent. Starting with state 1 we get $K_0 = \{1\}$, $K_1 = \{5\}$, and $K_2 = \{1\}$. Hence the irreducible closed set generated by state 1 is $\{1,5\}$ and has period 2. The set D corresponding to this closed set is $D = D_0 \cup D_1$ where $D_0 = \{1,3,7\}$ and $D_1 = \{3,5,7\}$. The new matrix Z^* is now of the form

$$Z^* = \begin{bmatrix} 0 & 0 & 0 & 0 & 0 & 0 & 0 & 0 & 0 & 0 \\ 0 & 0 & 0 & 0 & 0 & 0 & 0 & 0 & 0 & 0 \\ 0 & 0 & 0 & 0 & 0 & 0 & 0 & 0 & 0 & 0 \\ 0 & 0 & 0 & 0 & 0 & 0 & 0 & 0 & 0 & 0 \\ 0 & 0 & 0 & 0 & 0 & 0 & 0 & 0 & 0 & 0 \\ 0 & 0 & 0 & 0 & 0 & 1 & 0 & 0 & 0 & 0 \\ 0 & 0 & 0 & 0 & 0 & 0 & 0 & 0 & 0 & 0 \\ 0 & 0 & 0 & 0 & 0 & 0 & 0 & 0 & 0 & 0 \\ 0 & 0 & 0 & 0 & 0 & 0 & 0 & 0 & 0 & 0 \end{bmatrix}.$$

From this last matrix we see that state 6 is absorbing. Rewriting P in block form with the row and column ordering of $(2, 4, 8, 9, 1, 5, 6, 3, 7, 10)$ we get

$$
P = \begin{pmatrix}
\frac{1}{2} & \frac{1}{2} & 0 & 0 & 0 & 0 & 0 & 0 & 0 & 0 \\
0 & 0 & 1 & 0 & 0 & 0 & 0 & 0 & 0 & 0 \\
0 & 0 & 0 & 1 & 0 & 0 & 0 & 0 & 0 & 0 \\
1 & 0 & 0 & 0 & 0 & 0 & 0 & 0 & 0 & 0 \\
0 & 0 & 0 & 0 & 0 & 1 & 0 & 0 & 0 & 0 \\
0 & 0 & 0 & 0 & 1 & 0 & 0 & 0 & 0 & 0 \\
0 & 0 & 0 & 0 & 0 & 0 & 1 & 0 & 0 & 0 \\
0 & 0 & 0 & 0 & 0 & \frac{1}{2} & 0 & \frac{1}{2} & 0 & 0 \\
0 & 0 & 0 & 0 & \frac{1}{4} & \frac{1}{4} & \frac{1}{4} & \frac{1}{4} & 0 & 0 \\
\frac{1}{10} & \frac{1}{10} & \frac{1}{10} & \frac{1}{10} & \frac{1}{10} & \frac{1}{10} & \frac{1}{10} & \frac{1}{10} & \frac{1}{10} & \frac{1}{10}
\end{pmatrix}.
$$

The mean absorption times and absorption probabilities are found using the methods of Chapter III with

$$
Q = \begin{pmatrix}
\frac{1}{2} & 0 & 0 \\
\frac{1}{4} & 0 & 0 \\
\frac{1}{10} & \frac{1}{10} & \frac{1}{10}
\end{pmatrix} \quad \text{and} \quad R = \begin{pmatrix}
0 & 0 & 0 & 0 & 0 & \frac{1}{2} & 0 \\
0 & 0 & 0 & 0 & \frac{1}{4} & \frac{1}{4} & \frac{1}{4} \\
\frac{1}{10} & \frac{1}{10} & \frac{1}{10} & \frac{1}{10} & \frac{1}{10} & \frac{1}{10} & \frac{1}{10}
\end{pmatrix}.
$$

As stated previously the inversion of the matrix $I - Q$ on the computer can be done in many ways. This completes the analysis of this example.

We now turn to the analysis of an ergodic chain. In this case the main problem is to find the long run distribution. From Chapter III we know that if a stochastic matrix, P, is ergodic, the long run distribution is obtained by finding the solution of $\pi P = \pi$ and $\sum_{i=1}^{n} \pi_i = 1$. We will give two ways of solving this system on a computer.

Recall that the system $\pi P = \pi$ has infintely many solutions, so the second condition, $\sum_{i=1}^{n} \pi_i = 1$, must be used in order to get the long run distribution. One way of using this condition is to substitute the equation $\sum_{i=1}^{n} \pi_i = 1$ for any one of the equations $\pi_j = \sum_{k=1}^{n} \pi_k p_{kj}$ where $j = 1, 2, \ldots, n$. The resulting system of n equations in n unknowns will have a unique solution so this solution can be found using one of many standard programs.

The second approach to solving $\pi P = \pi$ and $\sum_{i=1}^{n} \pi_i = 1$ is to reduce the $n \times n$ system $\pi P = \pi$ to an $n - 1 \times n - 1$ system that has a unique solution and then use $\sum_{i=1}^{n} \pi_i = 1$ to get the long run distribution. That is, since the system $\pi P = \pi$ has one too many unknowns, by replacing one of the π_i's by the constant 1, we get a solvable system. The only problem with this approach is that the π_i which is set equal to one must not be a zero of the solution of $\pi P = \pi$ and $\sum_{i=1}^{n} \pi_i = 1$. In other words the coordinate, π_i, that is set equal to one must correspond to a positive persistent state, not a transient state. Therefore if this approach is to be used, a persistent state must be identifiable at the time ergodicity is determined.

The following lemmas show that we can be assured of finding a persistent state using the column sum method on an ergodic stochastic matrix. The first three lemmas show that the column sum method will always find a column whose sum is n or $n-1$ and the fourth lemma shows that this state is persistent.

Lemma VI.2.2. If P and Q are stochastic matrices and the kth column of P has no zeros, then the kth column of QP has no zeros.

Proof. See Exercise II.2.7. ▲

Lemma VI.2.3. Let P be an $n \times n$ transition matrix for a Markov chain ($n > 1$). If the chain is irreducible and aperiodic, then for each state i, there exists a $k \leqslant n - 1$ (where k may depend on i) such that $p_{ii}^{(k)} > 0$.

Proof. If for some state i, $p_{ii}^{(k)} = 0$ for $k = 1, 2, \ldots, n-1$, but $p_{ii}^{(n)} > 0$, then state i would be periodic of period n. (Why?) This contradicts the assumption of aperiodicity. ▲

It can be shown that if P is an $n \times n$ ergodic stochastic matrix then some power of P, say P^{k_0}, has a column with no zeros (Exercise 5). The next lemma provides an upper bound for the value of k_0.

Lemma VI.2.4. Let P be an $n \times n$ transition matrix for a Markov chain. Some power of P will have a column with no zeros if and only if $P^{(n-1)^2}$ does.

Proof. Assume some power of P has all positive entries in column j. Then state j corresponds to a persistent aperiodic state. Let C_j denote the closed set of states generated by the state j. The set C_j corresponds to the state space of an irreducible, aperiodic Markov chain with n or less states. By Lemma VI.2.3 there exists an integer $b \leqslant n - 1$ such that $p_{jj}^{(b)} > 0$. For notational convenience let us assume that $j = 1$. Let R denote the stochastic matrix P^b where we have $r_{11} > 0$. Using $R = \{r_{ij}\}$ as the new transition matrix it follows that $r_{i1}^{(n-1)} > 0$ for all states, i, in S. That is, since all the persistent states are assumed to intercommunicate with state 1 and since P is aperiodic, it must be possible to reach state 1 from all states using the transition matrix P^b. By Lemma VI.2.3 for each $i \neq 1$ there exists a non-negative $m_i \leqslant n - 1$ such that $r_{i1}^{(m_i)} > 0$. Now $r_{i1}^{(n-1)} \geqslant r_{i1}^{(m_i)} r_{11}^{(n-m_i-1)} > 0$.

Hence $R^{n-1} = P^{b(n-1)}$ has all positive entries in the first column. Hence by Lemma VI.2.2, $P^{(n-1)^2}$ has all positive entries in the first column.

The converse is logically true since if $P^{(n-1)^2}$ has a column with no zeros, then some power of P has a column with no zeros. ▲

In Section VI.4 we will consider some interesting consequences of Lemma VI.2.4.

Remark VI.2.1.   The above argument can be used to show that if P is an $n \times n$ stochastic matrix, then some power of P has no zeros if and only if $P^{n(n-1)}$ has no zeros (Exercise 7). The best result of this type is given in Perkins (1961) where he shows some power of P has no zeros if and only if P^{n^2-2n+2} does. (By the "best result" we mean that no smaller power of P can be used.)

Remark VI.2.2.   The best result concerning the power of P that has a column with no zeros is given in Isaacson and Madsen (1974). This theorem says that some power of an $n \times n$ transition matrix P will have a column of no zeros if and only if P^{n^2-3n+3} does. In view of this it is possible to give an improved version of the column sum method, so a slight improvement in the algorithm being given is possible. [That is, the upper bound of $(n-1)^2$ can be replaced by n^2-3n+3.] The justification of the bound n^2-3n+3 is again based on Perkins (1961) and does require some additional knowledge of algebra. Hence the more straightforward but slightly weaker development has been given here. The interested reader with a background in algebra will have little difficulty in reading the stronger form of Lemma VI.2.4 as given in Isaacson and Madsen (1974).

Let M be the power of Z at which the column sum method determines the corresponding Markov chain to be ergodic. Recall that the column sum method takes higher powers of Z until either some column of Z^M sums to n or $n-1$ and $\alpha(Z^M) > 0$ or until $M \geqslant (n-1)^2$. By Lemma VI.2.4 we see that if Z corresponds to an ergodic Markov chain, then the column sum method guarantees a column sum of at least $n-1$ in Z^M where $\alpha(Z^M) > 0$. If a column of Z^M sums to n, the state corresponding to this column is certainly persistent. The next lemma shows that if none of the columns of Z^M sum to n but $\alpha(Z^M) > 0$ and a column sums to $n-1$, then the state corresponding to that column is persistent.

Lemma VI.2.5.   Let P be an $n \times n$ transition matrix for a Markov chain. If the ith column of Z^M sums to $n-1$ and no column sums to n, and $\alpha(Z^M) > 0$ then state i is persistent.

Proof.   Assume there is a zero in the jth row and ones elsewhere in the ith column of Z^M. State i can definitely be reached from every state but j.

Hence the only way i can be transient is if j is absorbing. However, if state j is assumed to be absorbing, we will reach a contradiction. Specifically, if state j is absorbing, then the jth row of Z^M will have all zeros except in the (j,j) position, that is $z_{ji} = 0$ if $i \neq j$ and $z_{jj} = 1$. If we assume that the kth row has a zero in the jth column, then we have

$$\sum_{l=1}^{n} \min(z_{kl}, z_{jl}) = 0$$

or $\alpha(Z^M) = 0$, which contradicts the hypothesis of this lemma. On the other hand, if every row has a one in the jth column, then the jth column would sum to n. This also contradicts the hypothesis, hence we must conclude that state j is not absorbing and consequently that state i is persistent. ▲

In view of Lemma VI.2.5 the column sum method always yields a persistent state when ergodicity is determined. Hence the second method of solving $\pi P = \pi$ and $\sum_{i=1}^{n} \pi_i = 1$ can be used. That is, the π_i corresponding to this persistent state can be set equal to one. Any one of the n equations of $\pi P = \pi$ can be eliminated and the resulting $n-1 \times n-1$ system has a unique solution. Let π^* be this solution and consider $\pi^{**} = (\pi_1^{**}, \ldots, \pi_n^{**})$ where $\pi_j^{**} = \pi_j^*$, $j = 1, 2, \ldots, i-1$, $\pi_i^{**} = 1$, $\pi_j^{**} = \pi_{j-1}^*$, $j = i+1, \ldots, n$. The desired vector, π, is obtained from π^{**} by reducing the length of π^{**} to one unit. That is, let $\pi_i = \pi_i^{**} / \sum_{j=1}^{n} \pi_j^{**}$. Note that the $n-1 \times n-1$ system should be easier to solve than the $n \times n$ system. If some method other than the column sum method is used to determine ergodicity, the determination of a persistent state may not be so easy. In this case the first approach to solving $\pi P = \pi$ and $\sum_{i=1}^{n} \pi_i = 1$ should be used.

The final question to be considered for an ergodic stochastic matrix is that of the mean absorption times from the transient states. This is done in the same way it was done for the nonergodic matrices once the set of transient states is determined. At first glance this may seem to be an easy problem. That is, the transient states are those states i for which $\pi_i = 0$ in the long run distribution. This is true except there is a potential problem of round-off error in the computer. The value of π_i given by the computer may be essentially zero but not exactly zero. Hence to say that state i is transient if and only if $\pi_i = 0$ may lead to some errors. Various safeguards could be added to this approach, but each safeguard adds time to the analysis. We leave it as an exercise to discuss modifications of this approach that would be foolproof.

There is a second way to find the transient states that is foolproof. Unfortunately it usually is slower than simply looking at the long run distribution. If Z^{2^k} is calculated for $2^k > n(n-1)$, then a modification of Exercise 7 shows that a state is persistent if and only if the corresponding column of Z^{2^k} sums to n. (Again the result of Perkins can be used to show that $2^k > n^2 - 2n + 2$ is sufficient.) Unfortunately the column sum method usually determines the matrix to be ergodic long before $2^k \geqslant n(n-1)$ and provides only one persistent state, not all of them. Hence if this method of finding all persistent states were to be used, the column sum method should not be used. In this case powers of Z^{2^k} should be taken directly until $2^k \geqslant n(n-1)$. Then $\alpha(Z^{2^k})$ should be calculated and finally the analysis given.

The flow chart in Figure VI.2.1 describes the steps of the algorithm discussed in this chapter.

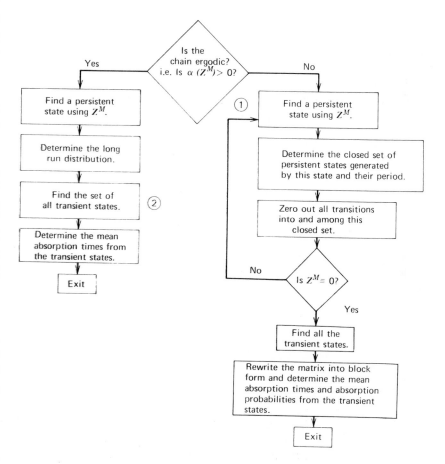

① Recall that $M \geqslant (n-1)^2$ so a persistent state is easily found.

② There are two ways of doing this. The programmer must make the choice between speed and accuracy.

Figure VI.2.1. Flow chart for analyzing a finite Markov chain.

The main purpose of this chapter has not been to motivate the reader to sit down and write a program using this algorithm. Rather it is hoped that he will consider various alternatives to this algorithm. In thinking about ways in which a stochastic matrix can be analyzed, the various characterizations of periodicity, irreducibility, and persistency become much clearer. In fact, new and better characterizations of these concepts might result from such considerations.

However, the algorithm does lead naturally into a program that will analyze finite Markov chains. In the next section we consider some of the possible uses of such a program.

EXERCISES

1. Using the adjacency matrices Z^{2^k} and the column sum approach, classify each of the states and express the transition matrix in block form, given that the transition matrix is

(a) $$P = \begin{bmatrix} \frac{1}{2} & 0 & \frac{1}{2} & 0 & 0 & 0 \\ 0 & 1 & 0 & 0 & 0 & 0 \\ \frac{1}{3} & 0 & \frac{2}{3} & 0 & 0 & 0 \\ 0 & 0 & 0 & 0 & 0 & 1 \\ \frac{1}{6} & \frac{1}{5} & \frac{1}{3} & \frac{1}{10} & \frac{1}{10} & \frac{1}{10} \\ 0 & 0 & 0 & 1 & 0 & 0 \end{bmatrix}$$

(b) $$P = \begin{bmatrix} 0 & \frac{1}{2} & 0 & \frac{1}{2} & 0 & 0 \\ 0 & 0 & \frac{1}{3} & 0 & \frac{2}{3} & 0 \\ 0 & 0 & 0 & \frac{1}{4} & \frac{3}{4} & 0 \\ 0 & 0 & 0 & 0 & 1 & 0 \\ \frac{1}{2} & 0 & 0 & \frac{1}{3} & 0 & \frac{1}{6} \\ 0 & \frac{1}{5} & \frac{4}{5} & 0 & 0 & 0 \end{bmatrix}.$$

2. If the transient states in the ergodic case are determined by looking at the long run distribution, what additional checks could be used to eliminate the possibility of errors due to round off?

3. Give an example of a stochastic matrix, P, such that $\alpha(Z) > 0$ and the column with the largest sum corresponds to a transient state.

*4. Show that if the matrix P is the transition matrix for an irreducible Markov chain, there is at least one positive element in every column.

*5. Prove that if P is an $n \times n$ ergodic stochastic matrix, then some power of P has a column with no zeros. (Hint: Recall that powers of P converge to a constant stochastic matrix.)

6. Prove the first claim in part 1 of the proof of Theorem VI.2.1. (Hint: Use Lemma III.1.1.)

7. Let P be an $n \times n$ stochastic matrix such that some power of P has no zeros. Show that $P^{n(n-1)}$ has no zeros. (This statement can be interpreted as follows: An $n \times n$ stochastic matrix is primitive if and only if $P^{n(n-1)}$ has no zeros.)

†8. Show that the criterion of maximum column sum can be applied to Z^* to find a persistent state. (See the discussion following Example VI.2.1 for a definition of Z^*.)

†9. Let K_0 be an irreducible aperiodic subset of the state space, $S = \{1, 2, \ldots, n\}$ for a nonergodic Markov chain. Choose $M \geqslant (n-1)^2$. Use Remark VI.2.1 to show that $p_{ij}^{(M)} > 0$ for all i and $j \in K_0$.

†10. Provide a detailed proof of part 2 of Theorem VI.2.1.

SECTION 3 **APPLICATIONS**

The main purpose of this chapter has been to present techniques that can readily be adapted for use in a computer program. That is, techniques that can be used to analyze finite Markov chains. We have been mainly concerned with techniques by which the Markov chains can be analyzed in a finite number of steps. Such techniques were described in the earlier sections of this chapter. In particular, in Section VI.1 an algorithm for determining ergodicity was described and a corresponding flow chart was shown in Fig. VI.1.1. In Section VI.2 an algorithm for determining the long run distribution and the set of transient states for ergodic chains and for determining closed sets and periodicity for nonergodic chains was described. The corresponding flow chart was shown in Fig. VI.2.1.

The two algorithms mentioned above form the basis for a computer program written by Isaacson and Sposito (1973). The program was constructed to be a general purpose program. It is admittedly inefficient if used to analyze a collection of Markov chains all of which are known to be either ergodic or nonergodic. In these less general cases it might be better to use a different program.

The following is a very brief description of the input parameters required from the user and the output options available using this program. In addition to the stochastic matrix itself, the input must include three additional parameters. The first parameter is the size of the matrix. The second parameter determines whether the chain should be analyzed or simply characterized as ergodic or nonergodic. The third parameter determines whether ergodicity should be determined using the column sum method or one of the other methods mentioned in Section VI.1. (For example: If the user strongly suspects the chain is nonergodic, the column sum method may not be the most efficient.)

The output for this program first reminds the user which type of analysis and method of analysis he requested. Then the stochastic matrix and corresponding adjacency matrix are given. Finally the matrix is stated to be ergodic or nonergodic and the appropriate analysis follows.

We now show how two Markov chains similar to chains described in Section III.6 would be analyzed using this program and exhibit parts of the output. We also discuss how this program could have been used on some other examples. Both ergodic and nonergodic matrices appeared in the applications of Chapter III. Hence we will analyze one of each type below.

By inspection of the first stochastic matrix given below it is clear that it is not ergodic since states 9 and 10 form a periodic closed subset. However, the answers to the following questions are not obvious.

   i)   Are there other closed subsets of this matrix and are they periodic?
  ii)   Is the second matrix given below ergodic?

The matrices were analyzed on an IBM 360-65 using a program based on the above algorithm. The total CPU time for these two examples was 3.42 seconds. The output from this analysis is given below.

ANALYSIS DESIRED:
   IF NONERGODIC, THEN GIVE A COMPLETE ANALYSIS
   IF ERGODIC, THEN DETERMINE THE LONG RUN
   DISTRIBUTION
THIS PROCEDURE USES THE COLUMN SUM METHOD
THE STOCHASTIC MATRIX IS

0.80	0.07	0.03	0.01	0.01	0.0	0.0	0.0	0.08	0.0
0.0	0.70	0.0	0.10	0.06	0.02	0.08	0.0	0.04	0.0
0.0	0.07	0.70	0.20	0.0	0.0	0.03	0.0	0.0	0.0
0.0	0.0	0.0	0.90	0.0	0.02	0.0	0.03	0.05	0.0
0.0	0.0	0.0	0.06	0.70	0.10	0.05	0.03	0.03	0.03
0.0	0.0	0.0	0.0	0.0	0.75	0.0	0.25	0.0	0.0
0.0	0.0	0.0	0.05	0.05	0.0	0.50	0.40	0.0	0.0
0.0	0.0	0.06	0.0	0.0	0.04	0.0	0.80	0.05	0.05
0.0	0.0	0.0	0.0	0.0	0.0	0.0	0.0	0.0	1.00
0.0	0.0	0.0	0.0	0.0	0.0	0.0	0.0	1.00	0.0

CHANGE ALL POSITIVE PROBABILITIES TO 1

1	1	1	1	1	0	0	0	1	0
0	1	0	1	1	1	1	0	1	0
0	1	1	1	0	0	1	0	0	0
0	0	0	1	0	1	0	1	1	0
0	0	0	1	1	1	1	1	1	1
0	0	0	0	0	1	0	1	0	0
0	0	0	1	1	0	1	1	0	0
0	0	1	0	0	1	0	1	1	1
0	0	0	0	0	0	0	0	0	1
0	0	0	0	0	0	0	0	1	0

THE STOCHASTIC MATRIX IS NONERGODIC
TO DETERMINE THE CLOSED SET W.R.T.   9

KO	9
K1	10
K2	9

PERIOD OF STATE 9 IS 2.

C1   9, 10

THE REMAINING STATES ARE TRANSIENT
IN ORDER TO REDUCE THE ORIGINAL STOCHASTIC MATRIX
TO BLOCK FORM, THE ROWS AND COLUMNS HAVE BEEN
REARRANGED WITH THE FOLLOWING ORDERING:

9   10   3   4   5   6   7   8   1   2

0.0	1.00	0.0	0.0	0.0	0.0	0.0	0.0	0.0	0.0
1.00	0.0	0.0	0.0	0.0	0.0	0.0	0.0	0.0	0.0
0.0	0.0	0.70	0.20	0.0	0.0	0.03	0.0	0.0	0.07
0.05	0.0	0.0	0.90	0.0	0.02	0.0	0.03	0.0	0.0
0.03	0.03	0.0	0.06	0.70	0.10	0.05	0.03	0.0	0.0
0.0	0.0	0.0	0.0	0.0	0.75	0.0	0.25	0.0	0.0
0.0	0.0	0.0	0.05	0.05	0.0	0.50	0.40	0.0	0.0
0.05	0.05	0.06	0.0	0.0	0.04	0.0	0.80	0.0	0.0
0.08	0.0	0.03	0.01	0.01	0.0	0.0	0.0	0.80	0.07
0.04	0.0	0.0	0.10	0.06	0.02	0.08	0.0	0.0	0.70

MEAN ABSORPTION TIME FOR EACH
OF THE TRANSIENT STATES:

3	21.999
4	18.550
5	18.104
6	19.500
7	18.065
8	15.500
1	16.872
2	19.255

PROB. OF ABSORPTION INTO THE CLOSED SET C(I)
FROM TRANSIENT STATE J IS GIVEN AS FOLLOWS:

TRANS. STATE	C(1)
3	1.000
4	1.000
5	1.000
6	1.000
7	1.000
8	1.000
1	1.000
2	1.000

The following is the analysis for a second stochastic matrix.

ANALYSIS DESIRED:
IF NONERGODIC, THEN GIVE A COMPLETE ANALYSIS
IF ERGODIC, THEN DETERMINE THE LONG RUN
DISTRIBUTION
THIS PROCEDURE USES THE COLUMN SUM METHOD
THE STOCHASTIC MATRIX IS

0.80	0.0	0.0	0.10	0.0	0.05	0.05	0.0
0.0	0.10	0.70	0.0	0.04	0.10	0.0	0.06
0.0	0.07	0.71	0.01	0.0	0.21	0.0	0.0
0.0	0.0	0.0	0.50	0.33	0.01	0.0	0.16
0.0	0.03	0.0	0.0	0.70	0.09	0.18	0.0
0.0	0.0	0.02	0.0	0.11	0.82	0.0	0.05
0.0	0.0	0.0	0.05	0.10	0.0	0.80	0.05
0.0	0.0	0.0	0.0	0.03	0.07	0.05	0.85

CHANGE ALL POSITIVE PROBABILITIES TO 1

1	0	0	1	0	1	1	0
0	1	1	0	1	1	0	1
0	1	1	1	0	1	0	0
0	0	0	1	1	1	0	1
0	1	0	0	1	1	1	0
0	0	1	0	1	1	0	1
0	0	0	1	1	0	1	1
0	0	0	0	1	1	1	1

THE STOCHASTIC MATRIX IS ERGODIC
THE LONG RUN PROBABILITY DISTRIBUTION IS:
$$0.0$$
$$0.010824$$
$$0.043150$$
$$0.025977$$
$$0.224041$$
$$0.246832$$
$$0.251144$$
$$0.198031$$

THE TRANSIENT STATES ARE:
   1
MEAN ABSORPTION TIME FOR EACH
OF THE TRANSIENT STATES
   1   5.00

The above output contains the basic information that is needed in the study of a finite Markov chain. This information could certainly be expanded or reduced by the programmer as he desired. The important fact about the program used here is that a single program analyzed both matrices, the ergodic and nonergodic. No preliminary analysis of the Markov chain by the user of the program is required.

In Application III.6.3, W. A. V. Clark studied the average rents of rental properties in central city tracts. For each city an 11×11 stochastic matrix was determined. If one were interested in analyzing many of these stochastic matrices, the program would be very useful, especially if some of the matrices happened to be nonergodic.

In Applications III.6.4 and III.6.5 the actual transition matrices were not given. However these two examples show that transition matrices are not always small in size. For large transition matrices the practice of analyzing Markov chains by inspection and/or trial and error becomes very difficult. Hence for these large stochastic matrices the program would again be useful.

In Application III.6.6 a stochastic matrix was constructed by using the "weight" assigned by each member of a group to the response (in this case a probability distribution) given by the members of the group. The weights, call them p_{ij}, were required to be non-negative and to satisfy $\Sigma_j \, p_{ij} = 1$. The value of p_{ij} then represents the "confidence" that individual i has in the response of individual j.

It is certainly not necessarily true that all such matrices would be ergodic. One might guess that in general p_{ii} would be positive and hence the matrices would be aperiodic; however, it is conceivable that, because of cliques or coalitions among individuals, the matrix could be reducible. If the number of individuals in the group were large and/or if there were a number of different matrices so constructed, it might be worthwhile to have the analysis of the matrices done on a computer.

It was stated at the beginning of this section that a program based on the algorithm of this chapter would be very inefficient if one wanted to calculate the long run distribution for 100 matrices that were known to be ergodic. This is because the first step of this algorithm is to determine ergodicity. In cases where something is known about the matrices to be analyzed, the reader might use parts of the above algorithm to write a smaller and faster program. For example, if a collection of stochastic matrices is known to be ergodic and only the long run distribution is desired, the method given in Paige, Styan, and Wachter (1973) is suggested.

SECTION **4** **INFINITE STOCHASTIC MATRICES**

In this final section of Chapter VI we digress from the general theme of analyzing Markov chains on a computer. Some of the lemmas and theorems developed in the first two sections of this chapter for the purpose of justifying the use of certain computer algorithms are of interest in their own right. We will discuss the extension of some of these lemmas and theorems to the case where the state space of the Markov chain is countably infinite. In the process we will discuss the general problem of extending results proven for a finite state space (or for finite matrices) to the case where the state space is countably infinite (or for infinite matrices).

For any statement that is true for finite Markov chains, it is logically true that one of the following cases holds:

(i)   The analagous statement for infinite chains is false;
(ii) The analagous statement for infinite chains is true and

$$\begin{matrix} \text{(a)} & \text{trivial to prove or} \\ \text{(b)} & \text{nontrivial to prove.} \end{matrix} \qquad (4.1)$$

[We would, by the way, certainly agree with the reader who asserts that the difference between ii(a) and ii(b) depends on your point of view!] In this section we will illustrate each of these cases.

We begin by giving some examples of statements that are true for finite stochastic matrices but false for infinite stochastic matrices.

Example VI.4.1. If P is a finite stochastic matrix with a column which contains no zeros, then $\alpha(P) > 0$ [or equivalently $\delta(P) < 1$]. The following example shows that this result need not be true for infinite stochastic matrices. (Why?)

$$P = \begin{pmatrix} \frac{1}{2} & \frac{1}{2} & 0 & 0 & 0 & \cdots \\ \frac{1}{3} & 0 & \frac{2}{3} & 0 & 0 & \cdots \\ \frac{1}{4} & 0 & 0 & \frac{3}{4} & 0 & \cdots \\ \frac{1}{5} & 0 & 0 & 0 & \frac{4}{5} & \cdots \\ \vdots & \vdots & \vdots & \vdots & \vdots & \end{pmatrix}.$$

Compare this example with Exercise VI.1.5.

Example VI.4.2. If P is a finite stochastic matrix such that $\lim_{n \to \infty} p_{ij}^{(n)} = \pi_j$ for all i and j, then P is weakly ergodic. For a counterexample in the infinite case, see Example V.3.4.

Example VI.4.3. If P is a finite stochastic matrix, then P has at least one invariant left probability vector. Exercise III.5.4 can be used to give a counterexample in the infinite case.

Example VI.4.4. It can be shown that if P is a finite stochastic matrix which has a positive column (i.e., a column which has no zero elements), then

$$\lim_{n \to \infty} \sup_{j} \sup_{i,k} |p_{ij}^{(n)} - p_{kj}^{(n)}| = 0 \qquad (4.2)$$

(Exercise 1). The following example shows that this result need not be true for infinite stochastic matrices. Let

$$P = \begin{bmatrix} \frac{1}{2} & \frac{1}{2} & 0 & 0 & 0 & \cdots \\ \frac{1}{4} & 0 & \frac{3}{4} & 0 & 0 & \cdots \\ \frac{1}{8} & 0 & 0 & \frac{7}{8} & 0 & \cdots \\ \frac{1}{16} & 0 & 0 & 0 & \frac{15}{16} & \cdots \\ \vdots & \vdots & \vdots & \vdots & \vdots & \end{bmatrix}.$$

It is easy to see that

$$p_{2,n+2}^{(n)} = \tfrac{3}{4} \cdot \tfrac{7}{8} \cdot \ \cdots\ \cdot \frac{2^n - 1}{2^n} = \prod_{k=2}^{n} (1 - 2^{-k})$$

and $p_{1,n+2}^{(n)} = 0$ for all n. From the theory of infinite products given in Chapter I, we know that $p_{2,n+2}^{(n)}$ decreases to some positive value, say d, as $n \to \infty$. Hence

$$\sup_{j} \sup_{i,k} |p_{ij}^{(n)} - p_{kj}^{(n)}| \geqslant |p_{2,n+2}^{(n)} - p_{1,n+2}^{(n)}| \geqslant d > 0.$$

We next consider some statements which were made in other sections of Chapter VI for finite chains, and for which the analagous statement for infinite chains is true and trivial to prove.

Example VI.4.5. In Exercise II.2.8 the reader was asked to show that if column j of P^m has no zeros, then state j is aperiodic. This clearly holds true if P is an infinite stochastic matrix.

Example VI.4.6. In Exercise VI.1.3 the reader was asked to show that if states i and j have a common consequent of order m, they have a common consequent of order $m+1$. This would clearly hold if P were an infinite stochastic matrix.

We next consider case (iib) of (4.1) where extensions to the infinite case hold but the proof is nontrivial.

Example VI.4.7. If P is a finite stochastic matrix corresponding to a weakly ergodic Markov chain, then it follows from Lemma VI.2.4 that some finite power of P has a column with no zeros. In fact the power required is decidable in terms of the size of P. The following theorem shows that a similar result is true for infinite stochastic matrices where the power is decidable in terms of the k and ϵ for which $\alpha(P^k) \geqslant \epsilon$.

Theorem VI.4.1.

If P is an infinite stochastic matrix such that $\alpha(P^k) > 0$ for some finite integer k, then some power of P has a column with no zeros.

Proof. Let $\epsilon = \alpha(P^k) > 0$. If we choose N such that $\Sigma_{j=N+1}^{\infty} p_{1j}^{(k)} \leqslant \epsilon/2$, then every row of P^k must have at least one positive element in the first N columns, for if not, then $\alpha(P^k)$ would be less than or equal to $\epsilon/2$. (Why?) Now let Q denote the $N \times N$ non-negative matrix in the upper left-hand corner of P^k. We know that since $\alpha(P^k) > 0$, it must be true that every pair of states must have a common consequent of order k and in particular this must be true for all pairs of states from the subset of states $\{1, 2, \ldots, N\}$. We would like to show that these pairs of states in fact have a common consequent of some order among the states $1, 2, \ldots, N$. It can be shown (Exercise 2) that every pair of states among the first N states has a common consequent among the first N states of order $2k$. Let Q^* denote the $N \times N$ matrix in the upper left-hand corner of P^{2k}. It follows from Exercise VI.2.5 that some power of Q^* has a column with no zeros, and in particular (from Lemma VI.2.4) that $(Q^*)^{(N-1)^2}$ has a column, say the j_0th column, with no zeros. From the definition of Q^*, we then know that $P^{2k(N-1)^2}$ has no zeros in the first N rows of the j_0th column, where $1 \leqslant j_0 \leqslant N$. Finally, since every row of P^k has a positive element in the first N columns, it follows that $P^k \cdot P^{2k(N-1)^2}$ has all positive values in the j_0th column. ▲

Remark VI.4.1. Since in the above proof we have the j_0th column of $(Q^*)^{(N-1)^2}$ with no zeros, it follows that $\xi = \min_{1 \leqslant i \leqslant N} p_{ij_0}^{2k(N-1)^2}$ is strictly positive, that is $\xi > 0$. It also follows from the way N was chosen that for all rows of P^k, $\Sigma_{j=1}^{N} p_{ij}^{(k)} \geqslant \epsilon/2$. (Why?) Hence every element of the j_0th column of $P^{2k(N-1)^2+k}$ must be at least as large as $\xi \cdot \epsilon/2 > 0$. In other words, not only are the elements of the j_0th column of $P^{2k(N-1)^2+k}$ positive, they are strictly bounded below away from zero. The fact that some finite power of a weakly ergodic stochastic matrix has a column with elements strictly bounded away from zero can be proved directly (Exercise 6).

Example VI.4.8. If P is a finite stochastic matrix for a weakly ergodic Markov chain, then P has a unique left invariant probability vector. In the following theorem we will show that the same result holds when P is infinite and weakly ergodic. However, we would emphasize that the assumption that P be weakly ergodic is crucial since the result may not be true for an arbitrary infinite stochastic matrix (see Example VI.4.3).

Theorem VI.4.2.

If P is an infinite stochastic matrix for a weakly ergodic Markov chain, then P has a unique left invariant probability vector.

Proof. A weakly ergodic chain has exactly one closed persistent subset (see Exercise V.3.4). Therefore the question concerning the existence of a left invariant probability vector reduces to the question of whether or not this persistent subset is null persistent (see Exercises III.5.3 and III.5.4).

It follows from Remark VI.4.1 that all elements in the j_0th column of P^m will be bounded below by $\xi \cdot \epsilon / 2$ whenever $m \geqslant 2k(N-1)^2 + k$ and $\alpha(P^k) = \epsilon$. Hence state j_0 is positive persistent (why?) so the one persistent subset corresponding to P must be positive persistent. Therefore there is a left invariant probability vector. It is left as an exercise to show that this left invariant probability vector is unique. ▲

Remark VI.4.2. In Remark V.4.2 we pointed out that the assumption in Corollary V.4.1 that P be in $\mathcal{C}$ is redundant, where $\mathcal{C}$ is defined in Definition V.4.1. This redundancy should now be clear.

We have given examples of results that hold true for finite Markov chains and which fall into the various categories given in (4.1). We have not attempted to give an exhaustive list, rather we have tried to alert the reader to the different possibilities that may occur when attempting to generalize results to infinite chains. We hope the reader will be at least a little suspicious of glib remarks to the effect that "the extension of these results to an infinite state space is obvious."

EXERCISES

1.   Prove that (4.2) holds for a finite stochastic matrix that has a column of no zeros.

2. In the proof of Theorem VI.4.1 show that every pair of states corresponding to the N rows of Q has a common consequent of order $2k$ within the first N states of S.

3. Show that a weakly ergodic stochastic matrix has a *unique* left invariant probability vector. The question here is the uniqueness.

4. Show that Theorem VI.4.2 can be proved in the following way. If P is weakly ergodic, then $\|P^n - Q\| \to 0$ as $n \to \infty$ where Q is stochastic. Hence a weakly ergodic stochastic matrix cannot be null persistent.

†**5.** Find a result that is true for finite Markov chains such that when it is extended to infinite chains, case i of (4.1) applies. Find examples for cases ii(a) and ii(b) also.

†**6.** Let P be a transition matrix for a weakly ergodic Markov chain. Show that some power of P has a column with elements strictly bounded away from zero.

CHAPTER VII

Continuous-Time Markov Chains

In the previous chapters we have restricted our attention to discrete-time Markov chains. As indicated in Chapter I, the extension to continuous state space requires mathematics well beyond the scope of this book. However, the extension of the time parameter from discrete time to continuous time is not difficult. The purpose of this chapter is to lay the foundations for the study of continuous-time Markov chains. We will not attempt to give a complete treatment of this subject. Rather we will give some basic theorems, definitions, and applications with emphasis given to how they relate to the discrete-time Markov chains.

SECTION 1 **DEFINITIONS**

Without loss of generality, assume the time parameter is the set of non-negative real numbers. The Markov property for a continuous-time Markov chain is defined as follows:

Definition VII.1.1. *A stochastic process*, $\{X_t\}_{t \in R^+}$, *is said to satisfy the* Markov property *if for all times* $t_0 < t_1 < \cdots < t_n < t$ *and for all n it is true that*

$$P\left[X_t = j \mid X_{t_0} = i_0, X_{t_1} = i_1, \ldots, X_{t_n} = i_n\right] = P\left[X_t = j \mid X_{t_n} = i_n\right]. \qquad (1.1)$$

As in the case of discrete time, this says that the present value of the process is all that counts in predicting the future. Notice, however, that for a discrete-time process the left-hand side of the equation for the Markov property considered all the past history. This is impossible for a continuous-time process since there are uncountably many time values in the past. This is the reason for requiring (1.1) to be true for all $t_0 < t_1 < \cdots < t_n < t$ and for all n.

By comparing the following definition with Definition I.3.2, we see that the concept of stationarity is essentially the same as in the discrete case.

Definition VII.1.2. *A Markov chain, $(X_t)_{t \in R^+}$, is said to be* stationary *if for every i and j the transition function, $P[X_{t+h} = j | X_t = i]$, is independent of t.*

For notational purposes we will denote the transition probabilities $P[X_{t+h} = j | X_t = i]$ by $p_{ij}(h)$ when the chain is stationary. As in the case of discrete-time stationary Markov chains, we can summarize the probabilities $p_{ij}(h)$ by using stochastic matrices $P(h)$, $h \geqslant 0$. For example, we might write $P(1) = (p_{ij}(1))$. We note, however, that in the case of continuous-time Markov chains there is no special significance to using the number 1, whereas in the discrete time case, one time unit was basic in the sense that transitions between states could only take place at times which were integer multiples of this time unit. In view of this time restriction in the discrete-time cases, the process could be described once $P(1)$ was known. In the continuous-time case $P(1)$ plays no such special role.

In order to describe a continuous-time stationary Markov chain using transition matrices, it is necessary to specify the entire family of stochastic matrices $\{P(t)\}_{t > 0}$. Obviously one can not use just any family of stochastic matrices, but only those families that satisfy certain conditions. For example, if the Markov property is to hold, then it must be that for all $s, t \in [0, \infty)$

$$p_{ij}(t+s) = \sum_{k \in S} p_{ik}(t) p_{kj}(s).$$

This equation is the continuous-time version of the Chapman–Kolmogorov equation. We will also assume that $p_{ij}(t)$ is continuous at zero. That is

$$\lim_{t \to 0^+} p_{ij}(t) = \begin{cases} 1 & \text{if} \quad i = j \\ 0 & \text{if} \quad i \neq j \end{cases}.$$

Note that this implies that $P(0)$ is the identity matrix.

For future reference we summarize the following conditions that must be satisfied by families of matrices used to describe continuous-time stationary Markov chains in this book.

Condition A. $p_{ij}(t) \geqslant 0$ for all $i, j \in S$ and $t \geqslant 0$.

Condition B. $\sum_{j \in S} p_{ij}(t) = 1$ for all $i \in S$ and $t \geqslant 0$.

Condition C. $p_{ij}(t+s) = \sum_{k \in S} p_{ik}(t) p_{kj}(s)$ for all $i, j \in S$ and $t, s \geqslant 0$.

Condition D. $\lim_{t \to 0^+} p_{ij}(t) = \begin{cases} 1 & \text{if} \quad i = j \\ 0 & \text{if} \quad i \neq j \end{cases}$.

(We would remind the reader that while the first three conditions are imposed to guarantee that the matrices are stochastic and the Markov property holds, we have imposed Condition D as a mathematical convenience.)

EXERCISES

1. Determine which of the following families of matrices satisfy conditions A–D of this section. That is, which have the properties of the transition probabilities for a continuous-time Markov chain?

(a) $P(t) = \begin{pmatrix} e^{-t} & 1 - e^{-t} \\ 0 & 1 \end{pmatrix}$ for $t \geqslant 0$.

(b) $P(t) = \begin{bmatrix} 1 - te^{-t} & te^{-t} & 0 & 0 \\ te^{-t} & 1 - 3te^{-t} & 2te^{-t} & 0 \\ 0 & te^{-t} & 1 - 2te^{-t} & te^{-t} \\ 0 & 0 & te^{-t} & 1 - te^{-t} \end{bmatrix}$ for $t \geqslant 0$.

(c) $P(t) = \begin{pmatrix} t + e^{-t} & 1 - t - e^{-t} \\ 0 & 1 \end{pmatrix}$ for $t \geqslant 0$.

(d) $P(t) = \begin{pmatrix} e^{t} & 1 - e^{t} \\ 0 & 1 \end{pmatrix}$ for $t \geqslant 0$.

(e) $P(t) = \begin{pmatrix} 1 & 0 \\ 1 - te^{-t} & te^{-t} \end{pmatrix}$ for $t \geqslant 0$.

SECTION 2   **PROPERTIES OF THE TRANSITION FUNCTION**

In this section we state and prove some theorems describing properties of the transition functions $p_{ij}(t)$ which follow from Conditions A–D of Section VII.1. Some of these theorems are of interest from a strictly mathematical viewpoint while other theorems are useful in describing some fundamental properties of continuous-time chains.

By imposing Condition D, we have required that $p_{ij}(t)$ be continuous at zero. The next theorem shows that in fact $p_{ij}(t)$ is a uniformly continuous function of t for $t \geqslant 0$.

Theorem VII.2.1.

Let X_t be a continuous-time Markov chain with transition functions $p_{ij}(t)$. If i and j are fixed states in S, then $p_{ij}(t)$ is a uniformly continuous function of t.

Proof. Let $\epsilon > 0$ be given. In order to show that $p_{ij}(t)$ is uniformly continuous, we need to show that there exists a $\delta > 0$ such that $|p_{ij}(t+h) - p_{ij}(t)| < \epsilon$ for all t whenever $0 < h < \delta$. Using the Chapman–Kolmogorov equation we have

$$|p_{ij}(t+h) - p_{ij}(t)| = \left| \sum_k p_{ik}(h) p_{kj}(t) - p_{ij}(t) \right|$$

$$\leqslant \sum_{k \neq i} p_{ik}(h) p_{kj}(t) + p_{ij}(t)|p_{ii}(h) - 1|. \qquad (2.1)$$

Since $p_{kj}(t) \leqslant 1$ for all k, j, and t it follows that

$$|p_{ij}(t+h) - p_{ij}(t)| \leqslant \sum_{k \neq i} p_{ik}(h) + |p_{ii}(h) - 1|.$$

But since $\sum_{k \in S} p_{ik}(h) = 1$, we have $\sum_{k \neq i} p_{ik}(h) = |p_{ii}(h) - 1|$. Hence

$$|p_{ij}(t+h) - p_{ij}(t)| \leqslant 2(1 - p_{ii}(h)),$$

and by Condition D we know that $\lim_{h \to 0^+} 1 - p_{ii}(h) = 0$. Therefore given $\epsilon > 0$ a $\delta > 0$ can be found so that $1 - p_{ii}(h) < \epsilon/2$ for all $0 < h < \delta$. Since this δ is independent of t, the proof is complete. (Actually the proof only establishes uniform continuity from the right. However it can be shown that this implies uniform continuity.) ▲

Having established the uniform continuity of $p_{ij}(t)$, the next logical question to consider is whether or not the function is differentiable. However, our reason for being interested in the derivative of $p_{ij}(t)$ is not simply academic. The derivatives of the functions $p_{ij}(t)$ for $i \in S, j \in S$, play a very important role in the study of continuous-time Markov chains. Before establishing the existence of these derivatives, we will give a heuristic discussion of why they are important.

In the discrete-time case the Markov chain was analyzed by simply looking at the one-step transition matrix. This matrix contained the transition probabilities of the next move. In the case of a continuous-time Markov chain the concept of the "next move" is not easily defined. Since the process is free to move continuously with time, there is no basic time unit $t^* > 0$ such that $p_{ij}(t^*)$ yields the transition probabilities of the next move. Since no positive time unit can be used in considering where the process will go next, we must restrict our attention to instantaneous changes. This leads directly to a consideration of the derivative of the functions $p_{ij}(t)$ for $i, j \in S$. We will see that even when the derivative exists it need not lie between zero and one, so it cannot be interpreted as a transition probability. We will return to this problem after establishing the differentiability of $p_{ij}(t)$.

Since the derivative is defined to be the limit of a difference quotient, we are again faced with the problem of determining whether a limit exists. In order to make this determination we will need to use the results of Lemma VII.2.1 below. The proof of this lemma involves certain properties of the supremum of a set of real numbers.

Lemma VII.2.1. Let $f(t)$ be a real-valued function defined on $(0, \infty)$. Assume that $\lim_{t \to 0^+} f(t) = 0$ and assume that f is subadditive [that is, $f(s + t) \leqslant f(s) + f(t)$ for all $s, t \in \mathbf{R}^+$]. Then $\lim_{t \to 0^+} [f(t)/t]$ exists and in fact equals $\sup_{t \in \mathbf{R}^+} [f(t)/t]$.

Proof. It is not assumed that $\sup_{t \in \mathbf{R}^+} [f(t)/t]$ is finite so the cases $\sup_{t \in \mathbf{R}^+} [f(t)/t] = b < \infty$ and $\sup_{t \in \mathbf{R}^+} [f(t)/t] = \infty$ will be considered separately.

First, assume that $\sup_{t \in \mathbf{R}^+} [f(t)/t] = b < \infty$. Let $\epsilon > 0$ be given. By the definition of supremum there exists $t^* > 0$ such that $f(t^*)/t^* > b - \epsilon$. Fix such a t^* and consider numbers t that are less than t^*. Dividing t into t^* we get

$$t^* = N_t \cdot t + r_t \tag{2.2}$$

where N_t is an integer depending on t and r_t is the remainder with $0 \leqslant r_t < t$. By applying the function f to both sides of (2.2), we get $f(t^*) = f(N_t \cdot t + r_t)$. Using the subadditivity of f it follows that

$$f(t^*) \leqslant f(N_t \cdot t) + f(r_t). \tag{2.3}$$

Since N_t is a positive integer, the subadditivity of f can be used again to yield

$$f(t^*) \leqslant N_t f(t) + f(r_t). \tag{2.4}$$

In order to show that $\lim_{t \to 0^+} [f(t)/t] = b$ it suffices to show that for t sufficiently small, $f(t)/t \geqslant b - 2\epsilon$. Dividing both sides of (2.4) by t^* we get

$$\frac{f(t^*)}{t^*} \leqslant \frac{N_t f(t)}{t^*} + \frac{f(r_t)}{t^*} = \frac{f(t)}{(t^*/N_t)} + \frac{f(r_t)}{t^*}. \tag{2.5}$$

Using (2.2) it follows that $t^*/N_t \geqslant t$, so

$$\frac{f(t^*)}{t^*} \leqslant \frac{f(t)}{t} + \frac{f(r_t)}{t^*}.$$

Since $0 \leqslant r_t < t$ and $\lim_{s \to 0^+} f(s) = 0$, a positive number T_1 can be chosen such that $f(r_t)/t^* < \epsilon$ for all $t \in (0, T_1)$. Hence for all $t \in (0, T_1)$ it follows that

$$\frac{f(t^*)}{t^*} - \epsilon < \frac{f(t)}{t}, \quad \text{so} \quad b - 2\epsilon < \frac{f(t)}{t}.$$

This completes the proof of the case $b < \infty$. The case where $b = \infty$ is handled in a similar manner and is left as an exercise. ▲

Theorem VII.2.2.

Let X_t be a continuous-time Markov chain with transition functions $p_{ij}(t)$. In the case where $i = j$ the resulting function of t has a right derivative at zero in the sense that $\lim_{t \to 0^+} [p_{ii}(t) - 1]/t = q_{ii}$ exists. This limit will always be nonpositive and in some cases the limit will be $-\infty$.

Proof. Condition C of Section VII.1 says that for all s, $t \in [0, \infty)$, $p_{ii}(t + s) = \sum_k p_{ik}(t) p_{ki}(s)$, so when the sum is reduced to the case where $k = i$ it follows that for all s, $t \in [0, \infty)$, $p_{ii}(t + s) \geqslant p_{ii}(t) p_{ii}(s)$. Consider the function $f(t) = -\log p_{ii}(t)$. It is easy to show that $f(t)$ is subadditive and the $\lim_{t \to 0^+} f(t) = 0$ (Exercise 2). By Lemma VII.2.1 it follows that $\lim_{t \to 0^+} [-\log p_{ii}(t)/t]$ exists. However, the difference quotient in question is $[p_{ii}(t) - 1]/t$, so we recall the following fact from calculus. For small values of $x > 0$, the Taylor's series for $-\log(1 - x)$ about zero is $-\log(1 - x) = x + x^2/2 + x^3/3 + x^4/4 + \cdots = x + R(x)$. The dominant term in this series is the first term. In fact, it can be shown that $\lim_{x \to 0^+} [R(x)/x] = 0$ (Exercise 3). Using the above facts we have

$$-\log p_{ii}(t) = -\log[1 - (1 - p_{ii}(t))] = 1 - p_{ii}(t) + R(1 - p_{ii}(t)).$$

Hence it follows that

$$\frac{\log p_{ii}(t)}{t} = \frac{p_{ii}(t) - 1}{t} - \frac{R(1 - p_{ii}(t))}{t}$$

$$= \frac{p_{ii}(t) - 1}{t}\left[1 + \frac{R(1 - p_{ii}(t))}{1 - p_{ii}(t)}\right]. \tag{2.6}$$

Therefore

$$\frac{p_{ii}(t) - 1}{t} = \frac{\log p_{ii}(t)}{t}\left[1 + \frac{R(1 - p_{ii}(t))}{1 - p_{ii}(t)}\right]^{-1}. \tag{2.7}$$

Taking the limits of both sides of (2.7) as $t \to 0^+$ and noting that both terms on the right-hand side have limits as $t \to 0^+$, we get that $\lim_{t \to 0^+} [p_{ii}(t) - 1]/t$ exists. ▲

The next theorem considers the right derivative of $p_{ij}(t)$ at zero when $i \neq j$.

Theorem VII.2.3

Let X_t be a continuous-time Markov chain with transition functions $p_{ij}(t)$. In the case where $i \neq j$ the resulting function of t has a right derivative at zero in the sense that $\lim_{t \to 0^+} [p_{ij}(t)/t] = q_{ij}$ exists and is finite.

Proof. In proving this theorem X_t will be considered only at certain discrete times. This is accomplished by only looking at the X_t process at times $h, 2h, 3h, \ldots$ where h is a small positive number. With this "discrete" version of X_t in mind, we note that $p_{ij}(mh) = p_{ij}^{(m)}(h)$. In other words, $p_{ij}(mh)$ will be viewed as an m-step transition of a discrete-time process with transition matrix, $p_{ij}(h)$. In the same way "taboo probabilities," $_jp_{ii}^{(m)}(h)$, are defined as follows:

$$_jp_{ii}^{(m)}(h) = P\big[X_{mh} = i, X_{(m-1)h} \neq j, \ldots X_h \neq j | X_0 = i \big].$$

That is, $_jp_{ii}^{(m)}(h)$ is the probability of going from state i to state i in m steps of size h without visiting j in the meantime. Using the arguments of Chapter II it follows that

$$p_{ij}(mh) \geqslant \sum_{k=0}^{m-1} {}_jp_{ii}^{(k)}(h) p_{ij}(h) p_{jj}(mh - kh - h). \tag{2.8}$$

[The right-hand side of (2.8) is smaller since not all possible transitions are considered.] By Condition D of Section VII.1 it follows that $\lim_{h \to 0^+} {}_jp_{ii}^{(k)}(h) = 1$. To see this write

$$p_{ii}^{(k)}(h) = {}_jp_{ii}^{(k)}(h) + \sum_{l=0}^{k} f_{ij}^{(l)}(h) p_{ji}\big[(k-l)h\big]$$

where $f_{ij}^{(l)}(h)$ is the probability of the first visit to j from i in l steps. Now

$$\lim_{h \to 0^+} p_{ji}\big[(k-l)h\big] = 0 \quad \text{so} \quad \lim_{h \to 0^+} {}_jp_{ii}^{(k)}(h) = \lim_{h \to 0^+} p_{ii}^{(k)}(h) = 1.$$

Let $\epsilon > 0$ be given and choose $t_0 > 0$ such that $p_{jj}(t) > 1 - \epsilon$ for all $t \in (0, t_0)$ and $_j p_{ii}^{(k)}(h) > 1 - \epsilon$ for all $kh \in (0, t_0)$. Then using (2.8) it follows that

$$p_{ij}(mh) \geqslant (1 - \epsilon)^2 \sum_{k=0}^{m-1} p_{ij}(h) = (1 - \epsilon)^2 m p_{ij}(h) \tag{2.9}$$

whenever $mh < t_0$. Divide (2.9) by mh and get

$$\frac{p_{ij}(mh)}{mh} \geqslant \frac{(1 - \epsilon)^2 p_{ij}(h)}{h} \quad \text{whenever } mh < t_0. \tag{2.10}$$

Let

$$L = \liminf_{t \to 0^+} \frac{p_{ij}(t)}{t}.$$

L is finite since, if m is chosen so that $t_0/2 \leqslant mh < t_0$, we have

$$\frac{p_{ij}(h)}{h} \leqslant \frac{p_{ij}(mh)}{mh} \cdot \frac{1}{(1 - \epsilon)^2} \leqslant \frac{1}{mh(1 - \epsilon)^2} \leqslant \frac{2}{t_0(1 - \epsilon)^2}.$$

Using the definition of liminf, choose $t_1 < t_0/2$ such that $p_{ij}(t_1)/t_1 < L + \epsilon$. By continuity there is an interval about t_1 where this inequality holds. That is, there exists $h_0 \in (0, t_1)$ such that $p_{ij}(t)/t < L + \epsilon$ for all t satisfying $|t - t_1| < h_0$. Choose $h < h_0$ and choose m so that

$$t_1 - h_0 < mh < t_1 + h_0 < t_0.$$

Using these choices of m and h and using (2.10) we have that

$$\frac{p_{ij}(h)}{h} \leqslant \frac{1}{(1 - \epsilon)^2} \frac{p_{ij}(mh)}{mh} \leqslant \frac{L + \epsilon}{(1 - \epsilon)^2}. \tag{2.11}$$

Since (2.11) holds for all $h < h_0$, it follows that

$$\limsup_{h \to 0^+} \frac{p_{ij}(h)}{h} \leqslant \frac{L + \epsilon}{(1 - \epsilon)^2}$$

but $\epsilon > 0$ is arbitrary so $\limsup_{h \to 0^+} [p_{ij}(h)/h] \leqslant L$. Therefore the limit of $p_{ij}(t)/t$ as $t \to 0^+$ exists and is finite. This limit will be denoted by q_{ij}. ▲

We now return to a discussion of what these derivatives mean. The quantity q_{ij} is non-negative but not necessarily between zero and one; hence we should not try to interpret it as a probability. However, we will see that q_{ij} is related to transitions from state i to state j and hence is called the *intensity of transition* from i to j. Similarly the quantity $-q_{ii}$ is nonnegative and we will see that it is related to the passage from state i to some other state. For this reason $-q_{ii}$ is called the *intensity of passage*. Specifically, we define q_{ii} by

$$-q_{ii} = \lim_{t \to 0^+} \frac{P[X_t \neq i | X_0 = i]}{t}.$$

The intensity of transition from i to j is related to transition probabilities in the following way. For small values of h, $P[X_{t+h} = j | X_t = i] = q_{ij} \cdot h + o(h)$ where $o(h)$ denotes a term that has the property that $\lim_{h \to 0^+}[o(h)/h] = 0$. Hence for small values of h, $q_{ij} \cdot h$ becomes a good approximation to the actual transition probability $p_{ij}(h)$. Without trying to decide just how small h must be before the approximation is good, we can definitely say that h must be small enough to make $q_{ij} \cdot h < 1$. Similarly $P[X_{t+h} \neq i | X_t = i] = P[X_h \neq i | X_0 = i] = -q_{ii} \cdot h + o(h)$. (The reader should recognize the approximation method used here as the method studied in introductory calculus. Namely if a function is differentiable at a point then one method of approximating the value of the function in a neighborhood of that point is to move along the tangent line.)

***Remark* VII.2.1.** It should be noted that the right-hand derivative of $P(t)$ does exist for all $t \geqslant 0$ but it was calculated at zero only. Actually for a stationary Markov chain the transition intensities are constant over time so it suffices to calculate the derivative at time zero. This is reminiscent of the stationary discrete-time case where only one transition matrix is needed for describing the behavior of the entire chain.

Since the matrix $Q = (q_{ij})$ is related to the transition probabilities for the "next move", Q often provides information that P provided in the discrete-time case. Before discussing this analogy, there is one important question that must be considered. First note that the intensity matrix, Q, of a continuous-time Markov chain with finite state space has the properties that $q_{ij} \geqslant 0$ and $q_{ii} = -\sum_{j \neq i} q_{ij}$ (Exercise 5). At this point we should ask whether every matrix Q with the property that $q_{ii} = -\sum_{j \neq i} q_{ij}$ determines a unique continuous-time Markov chain. Unfortunately the answer is no since there are cases where Q determines infinitely many transition matrix functions $P(t)$. [The interested reader is referred to a more advanced book e.g., Chung (1967) for details.] The following example shows one case where Q does determine $\{P(t)\}_{t \geqslant 0}$.

Example VII.2.1. When the state space S is finite, the matrices $\{P(t)\}_{t\geqslant0}$ can be determined from the intensity matrix Q. To see why this is true write the Chapman–Kolmogorov equation as

$$p_{ij}(t+\Delta t) = \sum_{k\in S} p_{ik}(t)p_{kj}(\Delta t). \qquad (2.12)$$

Next subtract $p_{ij}(t)$ from both sides of (2.12) and divide by Δt. The left-hand side of (2.12) becomes $[p_{ij}(t+\Delta t) - p_{ij}(t)]/\Delta t$, which can be shown to converge to $p'_{ij}(t)$ as $\Delta t \to 0$. The right-hand side of (2.12) becomes

$$\sum_{k\neq j} \frac{p_{ik}(t)p_{kj}(\Delta t)}{\Delta t} + p_{ij}(t)\left[\frac{p_{jj}(\Delta t)-1}{\Delta t}\right]. \qquad (2.13)$$

Taking the limit of (2.13) as $\Delta t \to 0$ we get $\sum_{k\neq j} p_{ik}(t)q_{kj} + p_{ij}(t)q_{jj}$. (Note that since the sum is finite, the interchange of limit and sum is easily justified.) Hence we get

$$p'_{ij}(t) = +p_{ij}(t)q_{jj} + \sum_{k\neq j} p_{ik}(t)q_{kj}.$$

In matrix notation this says

$$P'(t) = P(t)Q. \qquad (2.14)$$

A series solution to (2.14) is (Exercise 8)

$$P(t) = I + \sum_{n=1}^{\infty} \frac{Q^n t^n}{n!}. \qquad (2.15)$$

This series will converge and yield the unique solution $P(t)$ when Q is finite. Hence $\{P(t)\}_{t\geqslant0}$ can be determined from Q when Q is finite.

In the discrete-time case the matrix P was used to determine the long run distribution of the chain when the chain was ergodic. For continuous-time chains it is also true that if the long run distribution exists, it satisfies the system of equations

$$\pi_j = \sum_{i\in S} \pi_i p_{ij}(t) \quad \text{and} \quad \sum_{i\in S} \pi_i = 1 \quad \text{for all } j\in S \text{ and } t\in\mathbf{R}^+. \qquad (2.16)$$

If a Markov chain is defined by specifying Q, it may appear that the problem of finding the long run distribution is more difficult. However, in many cases this is not so. If we can differentiate the first equation of (2.16) with respect to t, we get a system of equations whose solution is the long run distribution. In performing this differentiation we are faced with the problem of interchanging the limit and the sum. If the state space is finite, there is no problem in justifying this interchange. However in general, conditions on

$$\lim_{t \to 0^+} \frac{p_{ij}(t)}{t} \quad \text{and} \quad \lim_{t \to 0^+} \frac{1 - p_{jj}(t)}{t}$$

must be given to allow for the interchange. Assuming the interchange is justified, differentiating both sides of (2.16) yields

$$0 = \sum_{i \neq j} \pi_i q_{ij} + \pi_j q_{jj} \quad \text{and} \quad \sum_{i \in S} \pi_i = 1 \quad \text{for } j \in S$$

so we get

$$-\pi_j q_{jj} = \sum_{i \neq j} \pi_i q_{ij} \quad \text{and} \quad \sum_{i \in S} \pi_i = 1 \quad \text{for } j \in S. \tag{2.17}$$

Hence we see that the matrix Q can sometimes be used to find the long run distribution of a continuous-time Markov chain and in this way provides some of the information that P provided in the discrete case. In particular, $\pi P(t) = \pi$ implies $\pi Q = \varphi$ so the system that is solved is $\pi Q = \varphi$ and $\sum_{i \in S} \pi_i = 1$.

We have seen how the matrix Q plays a role similar to the role of P in the case of discrete-time Markov chains. It would be ideal if one could say that the matrix Q describes the Markov chain in the same way that P does and that Q can always be used to find the long run distribution. Unfortunately this is not possible since, as stated previously, Q may not determine a unique Markov chain. However, in the remainder of this chapter we will restrict our attention to a special class of Markov chains where Q in fact does determine a unique chain. Hence these Markov chains will be described by giving Q rather than the family of stochastic matrices $\{P(t)\}_{t \geqslant 0}$.

EXERCISES

1. Prove Lemma VII.2.1 for the case $b = \infty$.
2. Show that the function $f(t) = -\log p_{ii}(t)$ is subadditive and that $\lim_{t \to 0^+} f(t) = 0$.

3. Show that

$$\lim_{x \to 0^+} \frac{1}{x}\left[\frac{x^2}{2} + \frac{x^3}{3} + \frac{x^4}{4} + \cdots\right] = 0.$$

4. Show that if the state space, X, of a continuous-time Markov chain is finite, then $-q_{ii} < \infty$ for all $i \in S$.

5. Show that if $P(t) = (p_{ij}(t))$ is the transition function for a continuous time finite Markov chain, then $q_{ii} = -\sum_{j \neq i} q_{ij}$ for all $i \in S$.

6. Let $Q = \begin{pmatrix} -1 & 1 \\ \frac{1}{2} & -\frac{1}{2} \end{pmatrix}$ be an intensity matrix for a continuous-time Markov chain. Find the transition matrices $\{P(t)\}$ associated with Q. Show that the matrices $\{P(t)\}$ satisfy Conditions A–D of Section VII.1.

7. Let $\{X_t\}$ be a Markov chain with intensity matrix

$$Q = \begin{bmatrix} -1 & 1 & 0 & 0 \\ 1 & -3 & 2 & 0 \\ 0 & 1 & -2 & 1 \\ 0 & 0 & 1 & -1 \end{bmatrix}$$

Find the long run distribution for this process.

8. Show that the series in (2.15) solves (2.14).

SECTION 3   **BIRTH AND DEATH PROCESSES**

As stated in Section VII.2, we now restrict attention to a special subclass of continuous-time stationary Markov chains. In particular we consider those continuous-time stationary Markov chains for which the state space is $S = \{0, 1, 2, \ldots\}$ and the transition intensities satisfy the condition

$$q_{ij} = 0 \quad \text{if} \quad |i - j| \geqslant 2. \tag{3.1}$$

Definition VII.3.1. *A continuous-time Markov chain on* $S = \{0, 1, 2, \ldots\}$ *that satisfies* (3.1) *is called a* birth and death process.

A birth and death process is a special type of counting process since $S = \{0, 1, 2, \ldots\}$ and hence the value of the process at time t represents the count at that time. It is also worth noting that condition (3.1) says that the movement of a birth and death process is up or down *one* unit at a time.

Since there are only two possible movements from state n, we define

$$q_{n,n+1} = \lambda_n$$

$$q_{n,n-1} = \mu_n$$

where $\mu_0 = \lambda_{-1} = 0$. From this it follows that $q_{nn} = -(\lambda_n + \mu_n)$.

Assume X_t is a birth and death process for which $0 < \lambda_n < \infty$ and $0 < \mu_{n+1} < \infty$ for $n = 0, 1, 2, \ldots, N$ where $N \leqslant \infty$. Then the corresponding Markov chain on $S = \{1, 2, \ldots, N\}$ will be irreducible and hence the limits

$$\lim_{t \to \infty} p_{ij}(t) = \pi_j \quad \text{for } j \in S$$

exist independently of $i \in S$. (For a proof of this fact the reader is referred to Chung, 1967.) (Note that periodicity plays no role in continuous time.) The question we want to consider is, how are the π_j's determined? In this special case we see that (2.17) reduces to

$$-\pi_n q_{nn} = \pi_{n-1} q_{n-1,n} + \pi_{n+1} q_{n+1,n}$$

since only two terms on the right-hand side of (2.17) are nonzero for a birth and death process. Using the λ_n's and μ_n's we get

$$\pi_n(\lambda_n + \mu_n) = \pi_{n-1}\lambda_{n-1} + \pi_{n+1}\mu_{n+1} \quad \text{for } n = 0, 1, 2, \ldots . \tag{3.2}$$

This simplifies to

$$\pi_n\lambda_n - \pi_{n+1}\mu_{n+1} = \pi_{n-1}\lambda_{n-1} - \pi_n\mu_n \quad \text{for } n = 0, 1, 2, \ldots . \tag{3.3}$$

When $n = 0$, the right-hand side of (3.3) is zero by the initial conditions, $\mu_0 = \lambda_{-1} = 0$. Hence $\pi_0\lambda_0 - \pi_1\mu_1 = 0$. Using (3.3) again for $n = 1$ it follows that $\pi_1\lambda_1 - \pi_2\mu_2 = 0$. Hence it follows by induction that $\pi_n\lambda_n - \pi_{n+1}\mu_{n+1} = 0$ for $n = 0, 1, 2, \ldots$. Therefore

$$\pi_{n+1} = \frac{\lambda_n}{\mu_{n+1}} \pi_n \quad \text{for } n = 0, 1, 2, 3, \ldots . \tag{3.4}$$

From (3.4) we have that

$$\pi_n = \frac{\lambda_{n-1}\lambda_{n-2}\cdots\lambda_0}{\mu_n\mu_{n-1}\cdots\mu_1} \pi_0,$$

so

$$1 = \sum_{k=0}^{\infty} \pi_k = \pi_0 \left[1 + \sum_{k=0}^{\infty} \frac{\lambda_k \lambda_{k-1} \cdots \lambda_0}{\mu_{k+1} \mu_k \cdots \mu_1} \right]. \tag{3.5}$$

If

$$\sum_{k=0}^{\infty} \frac{\lambda_k \lambda_{k-1} \cdots \lambda_0}{\mu_{k+1} \mu_k \cdots \mu_1} < \infty,$$

then

$$\pi_0 = \frac{1}{1 + \sum_{k=0}^{\infty} (\lambda_k \lambda_{k-1} \cdots \lambda_0 / \mu_{k+1} \mu_k \cdots \mu_1)}. \tag{3.6}$$

Using (3.6) and (3.4) all of the π_k's can be found. The π_k's represent the long run probability distribution of the process. We must ask, what happens if

$$\sum_{k=0}^{\infty} \frac{\lambda_k \lambda_{k-1} \cdots \lambda_0}{\mu_{k+1} \mu_k \cdots \mu_1} = \infty?$$

In this case $0 = \pi_0 = \pi_1 = \pi_2 = \cdots$ and hence $\{\pi_k\}$ is not a probability distribution. This says that there is zero probability of having any finite value for the process in the limit. Hence the value of the process will go to infinity as $t \to \infty$ with probability one. This situation is not new to us since we saw a similar behavior in the case of the irreducible null persistent chains of Chapter II. Hence we conclude that a birth and death process with intensities λ_n and μ_n has a long run probability distribution, $\{\pi_k\}$, if and only if

$$\sum_{k=0}^{\infty} \frac{\lambda_k \lambda_{k-1} \cdots \lambda_0}{\mu_{k+1} \mu_k \cdots \mu_1} < \infty.$$

If

$$\sum_{k=0}^{\infty} \frac{\lambda_k \lambda_{k-1} \cdots \lambda_0}{\mu_{k+1} \mu_k \cdots \mu_1} = \infty,$$

then $\pi_k = 0$ for $k = 0, 1, 2, \ldots$.

Example VII.3.1. Let X_t be a birth and death process for which $\lambda_n = 1$ for $n = 0, 1, 2, \ldots$ and $\mu_n = n$ for $n = 1, 2, 3, \ldots$. Find the long run probability distribution for X_t. In this case

$$1 + \sum_{k=0}^{\infty} \frac{\lambda_k \lambda_{k-1} \cdots \lambda_0}{\mu_{k+1} \mu_k \cdots \mu_1} = 1 + \sum_{k=0}^{\infty} \frac{1}{(k+1)!} = e.$$

Hence $\pi_0 = 1/e$, $\pi_1 = 1/e$, $\pi_2 = 1/(2e), \ldots, \pi_n = 1/(n!e), \ldots$.

Example VII.3.2. Let X_t be a birth and death process for which $\lambda_n = (n+1)\lambda$ and $\mu_n = n\mu$ for $n = 0, 1, 2, \ldots$. In order to find the long run distribution for X_t we need to evaluate $\sum_{k=0}^{\infty} (\lambda/\mu)^k$. This will be finite if and only if $\lambda < \mu$.

Example VII.3.2 has the following intuitive interpretation. Since $\lambda_n = (n+1)\lambda$ and $\mu_n = n\mu$, it is reasonable to say that the ratio of births to deaths is roughly λ/μ. That is if the population size is n, then $\lambda_n/\mu_n = (n+1)\lambda/n\mu$ which for large n is approximately λ/μ. Now if $\lambda/\mu \geqslant 1$ then no matter what the size of the population, there is a better chance of having a birth than a death and hence in the long run the population size will be infinite. A similar heuristic argument indicates that if $\lambda < \mu$ then the population size will stay finite.

Example VII.3.3. Let X_t be a birth and death process for which $\lambda_n > 0$ and $\mu_n = 0$ for $n = 0, 1, 2, \ldots$. A process for which $\mu_n \equiv 0$ is often called a *pure birth process*. There is little interest in the long run distribution of such processes since the population will go to infinity as $t \to \infty$ with probability one. However, various other properties of pure birth processes are of interest. In particular if $\lambda_n = \lambda$ for $n = 0, 1, 2, \ldots$ then the process is a standard Poisson process with parameter λ. This process is studied in great detail in most books on stochastic processes, so these results will not be repeated here. It is worth noting that the usual characterizations of the Poisson process do not explicitly involve the Markov property. In particular, one such characterization is given in terms of the distribution of the waiting times between occurrences. The reader interested in studying this and other characterizations of the Poisson process is referred to Parzen (1960).

Before turning to some applications of birth and death processes in queueing theory, we consider a second characterization of birth and death processes. Just as the Poisson process can be defined in terms of the distribution of the waiting times between occurrences of the events, the birth and death process can be defined in terms of the distributions of the waiting times until the next birth or death. In particular the following definition is equivalent to Definition VII.3.1.

Definition VII.3.2. $(X_t)_{t \geqslant 0}$ *is called a birth and death process if it satisfies the following conditions*:

i) X_t *is a counting process* (*that is,* $S = \{0, 1, 2, \dots\}$).
ii)   *When the chain is in state n, the waiting time until it moves to* $n + 1$ *has an exponential distribution with parameter* λ_n. [*Recall that such a distribution has density function* $f(x) = \lambda_n e^{-\lambda_n x}$ *for* $x > 0$ *and expected value equal to* $1/\lambda_n$.]
iii)   *When the chain is in state n, the waiting time until it moves to* $n - 1$ *has an exponential distribution with parameter* μ_n.
iv)   *These exponential distributions are independent.*

[See Ross (1972) for a development of birth and death processes using this definition.]

The parameters λ_n and μ_n in this definition correspond directly to the parameters λ_n and μ_n in Definition VII.3.1. To see why this should be true consider the following heuristic argument. In terms of Definition VII.3.1 λ_n denotes the intensity of passage from n to $n + 1$. Assume the units in which time is being measured are chosen so small that $\lambda_n < 1$. (For example, by reducing time from hours to seconds the corresponding value of λ_n could be reduced by the factor $1/3600$.) Under the assumption that $\lambda_n < 1$, we can use the intensity λ_n as an approximation to the probability of an arrival in the next time unit. Using this probability distribution and using the chosen time unit to define a discrete time process, the expected number of time units until the next occurrence is

$$\sum_{k=1}^{\infty} kP[\text{first occurrence at step } k]$$

$$= \sum_{k=1}^{\infty} k(1 - \lambda_n)^{k-1} \lambda_n = \lambda_n^{-1}.$$

Hence λ_n not only plays the role of the intensity of passage from n to $n + 1$ but λ_n also plays the role of the parameter of the exponential waiting time until the next arrival when the population size is n. In the same way it can be shown that μ_n denotes both the intensity of passage from n to $n - 1$ and the parameter of the exponential waiting time until the next death when the population size is n.

The following examples show how the birth and death processes arise in practice. These examples will be presented using the criterion of Definition VII.3.2 so the concept of a Markov chain will not be explicity used. However, in view of the equivalence of Definition VII.3.2 and Definition VII.3.1 the examples could be given as explicit applications of continuous time Markov chains. This interchangeability of Definitions VII.3.1 and VII.3.2 is very important in the study of birth and death processes. Hence even though we have not developed or discussed the waiting time approach, we hope the reader will recognize its usefulness.

Example VII.3.4. Assume customers arrive at a ticket counter with M windows according to a Poisson process with parameter 6 per minute. (This is equivalent to saying they arrive with independent exponential arrival times with parameter $\lambda = 6$.) Assume customers are served on a first-come–first-served basis. Assume that service times are independent and exponentially distributed with mean $\frac{1}{3}$ of a minute. Consider the following two questions related to this example:

1. What is the minimum number of windows needed to guarantee that the line does not get infinitely long?

2. Assume N_t denotes the number of customers waiting or being served at time t. Assume $M = 4$ and that a customer waits for service if $N_t \leqslant 4$, waits with probability $\frac{1}{2}$ if $N_t = 5$, and leaves if $N_t = 6$. What is the long run distribution for this process?

We first note that this example does describe a birth and death process with $\lambda_k = 6$ for $k = 0, 1, \ldots$ and $\mu_k = 3k$ for $k \leqslant M$ and $\mu_k = 3M$ for $k > M$. That μ_k assumes these values follows from the fact that the minimum of k independent exponential distributions is exponential with parameter equal to the sum of the respective parameters (Exercise 1). It follows that

$$\sum_{k=0}^{\infty} \frac{\lambda_k \lambda_{k-1} \cdots \lambda_0}{\mu_{k+1} \mu_k \cdots \mu_1} < \infty \tag{3.7}$$

if $M \geqslant 3$ since in this case

$$\frac{\lambda_k}{\mu_{k+1}} = \frac{6}{3M} < 1 \quad \text{for } k \geqslant M - 1.$$

For the question in part 2 the λ_k's are changed by the fact that customers do not stay if the counter is crowded. This means that $\lambda_k = 0$ for $k \geqslant 6$ and hence (3.7) becomes a finite series. Under the assumptions of part 2 we have $\lambda_0 = \lambda_1 = \lambda_2 = \lambda_3 = \lambda_4 = 6$ and $\lambda_5 = 3$. Hence

$$\pi_0 + \pi_0 \left[\sum_{k=0}^{5} \frac{\lambda_k \lambda_{k-1} \cdots \lambda_0}{\mu_{k+1} \mu_k \cdots \mu_1} \right] = 1$$

so

$$\pi_0 = \frac{12}{89}, \qquad \pi_1 = \frac{24}{89}, \qquad \pi_2 = \frac{24}{89}, \qquad \pi_3 = \frac{16}{89},$$

$$\pi_4 = \frac{8}{89}, \qquad \pi_5 = \frac{4}{89}, \quad \text{and} \quad \pi_6 = \frac{1}{89}.$$

Example VII.3.5. A taxi company has one mechanic who replaces fuel pumps when they fail. Assume the waiting time in days until a fuel pump fails is exponentially distributed with parameter $1/300$. (Hence the expected time until the next failure for each car is 300 days.) Assume the company has 1000 cars. Assume the repair time for each car is exponentially distributed with expected repair time of $\frac{1}{4}$ days. Find the long run distribution for the process X_t, where $X_t =$ number of cars with a broken fuel pump at time t.

For this process a birth corresponds to a broken fuel pump and a death corresponds to a repaired pump. Now $\lambda_n = (1000 - n)/300$, and $\mu_{n+1} = 4$ for $n = 0, 1, 2, \ldots, 1000$. At this point the arithmetic becomes difficult but it is easy to see that the long run distribution will exist since $\lambda_{1000} = 0$. That is, there can never be more than 1000 cars with broken fuel pumps. In particular

$$\pi_0 = \frac{1}{1 + \displaystyle\sum_{k=0}^{999} \frac{(1000)(999) \ldots (1000 - k)}{(1200)^{k+1}}}$$

and the other π_i's are determined from π_0.

There are many other areas in which the birth and death process model is used. There are also many other interesting questions in addition to the long run distribution that can be considered. However, these questions are considered in most texts on stochastic processes and will not be repeated here.

As stated in Section VII.1 the purpose of this chapter has been to give the basic results of continuous-time Markov chains and relate them to the discrete-time case. In this section a brief introduction to the uses of the continuous time Markov chains has been given. In order to go further in this area a better understanding of stochastic processes is helpful. Since the main thrust of this book is on Markov chains with emphasis on discrete time, we will leave the topic of birth and death processes here. However, we wish to make it clear that much more can and has been said on the topic and that an interested reader will have little difficulty in finding textbooks that say it well.

EXERCISES

1. Let X_1 be exponential with parameter λ_1. Let X_2 be exponential with parameter λ_2. Show that if X_1 and X_2 are independent, then $\min(X_1, X_2)$ is exponential with parameter $\lambda_1 + \lambda_2$.

2. In Example VII.3.4 find the long run distribution if there are four ticket windows and customers wait for service no matter how long the lines are.

†3. Let $Z(t)$ be a pure birth process for which $P[Z(t+h) = i+1 | Z(t) = i] = \lambda_i h + o(h)$ and $P[Z(t+h) \geq i+2 | Z(t) = i] = o(h)$. Let $P_n(t) = P[Z(t) = n]$. Show that

   i) $P_0'(t) = -\lambda_0 P_0(t)$

   ii) $P_n'(t) = -\lambda_n P_n(t) + \lambda_{n-1} P_{n-1}(t)$ for $n \geq 1$.

   Solve this system of equations in the special case where $\lambda_n \equiv \lambda$ for all n and $P_0(0) = 1$. In this case $Z(t)$ is a Poisson process.

4. Consider a barber shop with two barbers. Assume their service times are independent and exponentially distributed with mean 20 minutes ($\mu = 1/20$) when there are at most two customers in the shop. This reduces to 15 minutes if there are three or more customers in the shop. Assume customers arrive according to a Poisson process with parameter $\lambda = 1/10$ when there are two or less customers in the shop. This reduces to $\lambda = 1/30$ if there are three or more customers in the shop. Let X_t denote the number of customers in the shop at time t. Find the long run distribution for X_t. What is the long run probability that a customer must wait for service?

5. Find the long run distribution for the $X(t)$ in Example VII.3.5 assuming the taxi company has only eight taxis. What is the long run distribution for $X(t)$ if a second mechanic is hired? (Assume the second mechanic works at the same rate as the first.)

6. Cars arrive at a service station according to a Poisson process with parameter $\lambda = 30$ per hour. Service times for these cars are indepen-

dently and exponentially distributed with mean 5 minutes. This particular station has three attendants who all work at the same rate. You want to take your car to this station late in the afternoon but you do not want to wait for service. What is the probability that at least one attendant will be available when you arrive. (Assume that the long run distribution is in effect late in the afternoon.) Across town there is a smaller one attendant station at which cars arrive according to a Poisson process with parameter $\lambda = 8$ per hour. The attendant works at the same rate as those at the larger station. Would a customer be better off going to this station?

References

Anderson, Subra (1970). *Graph Theory and Finite Combinatorics*. Markham Publishing, Chicago.

Ash, R. and R. Bishop (1972). "Monopoly as a Markov process." *Mathematics Magazine*, **45**, pp. 26–29.

Birkhoff, G. and S. MacLane (1953), *A Survey of Modern Algebra*. McMillan, New York.

Blumen, I., M. Kogan, and P. J. McCarthy (1955). "The industrial mobility of labor as a probability process." *Cornell Studies of Industrial and Labor Relations*, Vol. 6. Ithaca, New York.

Bowerman, Bruce (1974). "Nonstationary Markov decision processes and related topics in nonstationary Markov chains," Ph.D. thesis, Iowa State University.

Bowerman, B., H. T. David, and D. Isaacson (1974). "Uniformly Strong Ergodicities of Markov Chains." Technical Report, Statistical Laboratory, Iowa State University.

Chung, K. L. (1967). *Markov Chains with Stationary Transition Probabilities*, 2nd ed., Springer-Verlag, Berlin.

Clark, W. A. V. (1965). "An application to the movement of rental housing areas." *Association of American Geographers Annals*, **55**, pp. 351–359.

Degroot, M. H. (1974). "Reaching a Consensus." *Journal of the American Statistical Association*, **69**, pp. 118–121.

dePillis, John (1969). *Linear Algebra*. Holt-Rinehart-Winston, New York.

Dobrushin, R. L. (1956). "Central limit theorems for non-stationary Markov chains II." *Theory of Probability and its Applications*, **1**, pp. 329–383 (English translation).

Fadeev, D. K. and V. N. Fadeeva (1963). *Computational Methods of Linear Algebra*. W. H. Freeman, San Francisco, Calif.

Feller, William (1968). *An Introduction to Probability Theory and its Applications*, 3rd ed. Wiley, New York.

Fox, B. L. and D. M. Landi (1968). "An algorithm for identifying the ergodic subchains and transient states of a stochastic matrix." *Communication of the Association for Computing Machinery* **11**, pp. 619–621.

Gantmacher, F. R. (1959). *Applications of the Theory of Matrices*. Wiley-Interscience, New York.

Griffeath, D. (1975). "Uniform coupling of nonhomogeneous Markov chains." *Journal of Applied Probability* **12**.

Hajnal, J. (1956). "The ergodic properties of non-homogeneous finite Markov chains." *Proceedings of the Cambridge Philosophical Society*, **52**, pp. 67–77.

Hajnal, J. (1958). "Weak ergodicity in nonhomogeneous Markov chains." *Proceedings of the Cambridge Philosophical Society*, **54**, pp. 233–246.

Henry, N., R. McGinnis, and H. Tegtmeyer (1971). "A finite model of mobility." *Journal of Mathematical Sociology*, 1, pp. 107–118.

Hentzel, I. (1973). "How to win at monopoly." *Saturday Review of the Sciences*, pp. 44–48.

Hoffman, K. and R. Kunze (1962). *Linear Algebra*. Prentice-Hall, Englewood Cliffs, New Jersey.

Iosifescu, M. and R. Theodorescu (1969). *Random Processes and Learning*. Springer-Verlag, Berlin.

Iosifescu, M. (1972). "On two recent papers on ergodicity in nonhomogeneous Markov chains." *Annals of Mathematical Statistics*, 43, pp. 1732–1737.

Isaacson, D. and R. Madsen (1974). "Positive columns for stochastic matrices." *Journal of Applied Probability*, 11, pp. 829–835.

Isaacson, D. and V. Sposito (1973). "GAUR, A computational procedure to analyze a Markov chain." Numerical Analysis-Programming Series No. 15, Statistical Laboratory, Iowa State University, Ames, Iowa.

Karlin, Samuel (1968). *A First Course in Stochastic Processes*. Academic Press, New York.

Kemeny, G. and J. L. Snell (1960). *Finite Markov Chains*. D. Van Nostrand, Princeton, New Jersey.

Krenz, R. D. (1964). "Projection of farm numbers for North Dakota with Markov chains." *Agricultural Economics Research*, XVI, No. 3, p. 77.

Lee, K. H. (1968). "A Markov chain analysis of caries process with consideration for the effect of restoration." *Archives of Oral Biology*, 13, pp. 1119–1132.

Madsen, R. and D. Isaacson (1973). "Strongly ergodic behavior for nonstationary Markov processes," *Annals of Probability*, 1, pp. 329–335.

Madsen, R. (1975). "Decidability of $\alpha(P^k) > 0$ for some k." *Journal of Applied Probability*, 12, pp. 333–340.

McGinnis, R. (1968). "Stochastic model of social mobility." *American Sociological Review*, 33, pp. 712–722.

Meyer, Paul L. (1970). *Introductory Probability and Statistical Applications*, 2nd ed. Addison-Wesley, Reading, Massachusetts.

Mott, J. L. (1957). "Conditions for the ergodicity of non-homogeneous finite Markov chains." *Proceedings of the Royal Society of Edinburgh*, 64, pp. 369–380.

Orey, S. (1961). "Strong ratio limit property." *Bulletin of the American Mathematical Society*, 67, pp. 571–574.

Paige, C. C., G. P. H. Styan, and P. G. Wachter (1973). "Computation of the stationary distribution of a Markov chain." Reprint from McGill University.

Parzen, E. (1967). *Stochastic Processes*. Holden-Day Inc., San Francisco, California.

Paz, A. (1963). "Graph-theoretic and algebraic characterizations of some Markov Processes." *Israel Journal of Mathematics*, 3, pp. 169–180.

Paz, A. (1970). "Ergodic theorems for infinite probabilistic tables." *Annals of Mathematical Statistics*, 41, pp. 539–550.

Paz, A. (1971). *Introduction to Probabilistic Automata*. Academic Press, New York.

Perkins, Peter (1961). "A theorem on regular matrices." *Pacific Journal of Mathematics*, 11, pp. 1529–1533.

Pitman, J. W. (1974). "Uniform rates of convergence for Markov chain transition probabilities." *Zeitschrift fur Wahrscheinlichkeitstheorie verwandte und Gebiete*, 29, pp. 193–227.

Rao, C. R. (1965). *Linear Statistical Inference and its Applications*. Wiley, New York.

Ross, S. (1970). *Applied Probability Models with Optimization Applications*. Holden-Day, San Francisco, Calif.

Ross, S. (1972). *Introduction to Probability Models*. Academic Press, New York.

Seneta, E. (1973). *Non-Negative Matrices*. Halsted Press, New York.

Westlake, J. R. (1968). *A Handbook of Numerical Matrix Inversion and Solution of Linear Systems*. Wiley, New York.

Index